D0502524

A REASONABLE DEFENSE

Studies in Defense Policy
SELECTED TITLES IN PRINT

U.S. Tactical Air Power: Missions, Forces, and Costs
William D. White

U.S. Nuclear Weapons in Europe: Issues and Alternatives
Jeffrey Record with the assistance of Thomas I. Anderson

The Control of Naval Armaments: Prospects and Possibilities
Barry M. Blechman

Stresses in U.S.-Japanese Security Relations
Fred Greene

The Military Pay Muddle
Martin Binkin

Sizing Up the Soviet Army
Jeffrey Record

Where Does the Marine Corps Go from Here?
Martin Binkin and Jeffrey Record

Deterrence and Defense in Korea: The Role of U.S. Forces
Ralph N. Clough

Women and the Military
Martin Binkin and Shirley J. Bach

The Soviet Military Buildup and U.S. Defense Spending
Barry M. Blechman and others

Soviet Air Power in Transition
Robert P. Berman

Shaping the Defense Civilian Work Force: Economics, Politics, and National Security
Martin Binkin with Herschel Kanter and Rolf H. Clark

The Military Equation in Northeast Asia
Stuart E. Johnson with Joseph A. Yager

Youth or Experience? Manning the Modern Military
Martin Binkin and Irene Kyriakopoulos

Paying the Modern Military
Martin Binkin and Irene Kyriakopoulos

Defense in the 1980s
William W. Kaufmann

The FX Decision: "Another Crucial Moment" in U.S.-China-Taiwan Relations
A. Doak Barnett

Planning Conventional Forces, 1950–80
William W. Kaufmann

Blacks and the Military
Martin Binkin and Mark J. Eitelberg with Alvin J. Schexnider and Martin M. Smith

U.S. Arms Sales: The China-Taiwan Tangle
A. Doak Barnett

Soviet Strategic Forces: Requirements and Responses
Robert P. Berman and John C. Baker

U.S. Ground Forces and the Defense of Central Europe
William P. Mako

Alliance Security: NATO and the No-First-Use Question
John D. Steinbruner and Leon V. Sigal, Editors

The 1985 Defense Budget
William W. Kaufmann

America's Volunteer Military: Progress and Prospects
Martin Binkin

The 1986 Defense Budget
William W. Kaufmann

The Calculus of Conventional War: Dynamic Analysis without Lanchester Theory
Joshua M. Epstein

A Reasonable Defense
William W. Kaufmann

STUDIES IN DEFENSE POLICY

A REASONABLE DEFENSE

William W. Kaufmann

Library of Congress Catalog Card Number 85-73331

ISBN 0-8157-4879-5

9 8 7 6 5 4 3 2 1

THE BROOKINGS INSTITUTION is an independent organization devoted to nonpartisan research, education, and publication in economics, government, foreign policy, and the social sciences generally. Its principal purposes are to aid in the development of sound public policies and to promote public understanding of issues of national importance.

The Institution was founded on December 8, 1927, to merge the activities of the Institute for Government Research, founded in 1916, the Institute of Economics, founded in 1922, and the Robert Brookings Graduate School of Economics and Government, founded in 1924.

The Board of Trustees is responsible for the general administration of the Institution, while the immediate direction of the policies, program, and staff is vested in the President, assisted by an advisory committee of the officers and staff. The by-laws of the Institution state: "It is the function of the Trustees to make possible the conduct of scientific research, and publication, under the most favorable conditions, and to safeguard the independence of the research staff in the pursuit of their studies and in the publication of the results of such studies. It is not a part of their function to determine, control, or influence the conduct of particular investigations or the conclusions reached."

The President bears final responsibility for the decision to publish a manuscript as a Brookings book. In reaching his judgment on the competence, accuracy, and objectivity of each study, the President is advised by the director of the appropriate research program and weighs the views of a panel of expert outside readers who report to him in confidence on the quality of the work. Publication of a work signifies that it is deemed a competent treatment worthy of public consideration but does not imply endorsement of conclusions or recommendations.

The Institution maintains its position of neutrality on issues of public policy in order to safeguard the intellectual freedom of the staff. Hence interpretations or conclusions in Brookings publications should be understood to be solely those of the authors and should not be attributed to the Institution, to its trustees, officers, or other staff members, or to the organizations that support its research.

FOREWORD

As the debate over reducing the federal deficit has intensified, questions about the U.S. defense program and budget have proliferated. Should the defense budget be frozen for one year or for several? What should be done about the unspent balances of budget authority approved in prior years? If resources for defense are to be constrained, which programs should be cut? Does the country need a new approach to defense or should it continue to tinker with the old?

The stakes are high, the issues are complex, and feelings run strong. Yet there is no reason why defense should be any less subject to rational discourse than other government programs. The language of defense budgeting, the processes used to develop the defense budget, and the amenability to control of past, current, and future defense budgets are not beyond the grasp of interested laymen. The trends in national defense spending and the continuity of policy over the last forty years can be readily determined and used as a basis for judging the need for change. The arguments for and against restraint in defense spending can be evaluated. And it is possible to examine systematically the return in security the country can expect on its programmed defense investment, compare that return with alternative investments, ask why the Pentagon has not chosen a more efficient mix of capabilities, and consider what can be done to create more cost-effective defenses and to help reduce the federal deficit.

In this study William W. Kaufmann comments on all these subjects. However, because cost-effectiveness of defense investment has become such a salient issue, he concentrates on assessing how three different U.S. force structures would respond to a number of standard planning contingencies. He compares the relative effectiveness of these force structures in countering a Soviet first strike with strategic nuclear forces and a nonnuclear invasion of Central Europe by the Warsaw Pact. He finds that if U.S. forces could be coherently designed to address major U.S. vulnerabilities, they would not only outperform the currently programmed force, but would save at least $200 billion in outlays between fiscal 1986 and fiscal 1990. This finding admittedly comes from comparing an ideal force with one that results from the political process (both inside and outside the Pentagon). Its thrust, nonetheless, is that the national interest would be better served by a full-scale review of U.S. strategy, forces, and budgets than by marginal adjustments to the current defense program.

William W. Kaufmann is a consultant to the Brookings Foreign Policy Studies program and a member of the faculty of the John F. Kennedy School of Government at Harvard University. He is grateful to Robert Howard, Robert Komer, Philip Odeen, Joseph A. Pechman, and John D. Steinbruner for their comments on the manuscript. Caroline Lalire and Theresa B. Walker edited the manuscript, James E. McKee verified its factual content, and Gregory K. Tally did the proofreading. Kathryn Ho contributed unequaled secretarial and administrative skills. Funding for this study was provided in part by the Ford Foundation and by the John D. and Catherine T. MacArthur Foundation.

BRUCE K. MAC LAURY
President

January 1986
Washington, D.C.

CONTENTS

1. **Disarray in Defense** 1

2. **Controlling the Defense Budget** 6

   Definitions of Defense—Budget Authority—Balances of Budget Authority—Budget Formats—The Budget Cycle—The Five-Year Defense Program

3. **Trends in National Defense** 18

   Isolation—The Truman Years—From Eisenhower to Reagan—The Reagan Revolution—Goals

4. **Debating Points and Force Planning** 31

   Defense and the Deficit—Waste, Fraud, and Abuse—The Threat—The Decade of Neglect—The Window of Vulnerability—Return on Investment—Force Comparisons—Military Experience and Judgment

5. **Net Assessments and Force Planning** 58

   Forces, Costs, and Performance—Naval Forces—The Three Forces—Performance of the Forces—The Reasons for the Differences

6. **What Has Gone Wrong** 93

   The Role of the Secretary of Defense—Planning, Programming, Budgeting—Centralized Planning—The Decline of the Planning-Programming-Budgeting System—Inefficiencies—Exaggerated Threats—The Congressional Role

7. **What Can Be Done** 106

   The Competition—Future Growth—The Defense Intelligence Agency—Net Assessment—Policy—Analysis—The Joint Chiefs of Staff and the Secretary of Defense

Tables

1-1. Defense Budget Estimates, Fiscal Year 1986 3
1-2. Five-Year and Three-Year Defense Plans, Fiscal Years 1986–90 3
1-3. Deficit Estimates, Fiscal Years 1986–90 4
2-1. Budget Authority and Outlays by Function, Fiscal Years 1984–86 7
2-2. Outlays, by Appropriation Categories and Major Weapons, as a Percent of First-Year Budget Authority Spent Each Year over a Six-Year Period 11
2-3. Balances of Prior-Year Budget Authority for Defense, End of Fiscal Years, 1980–86 11
2-4. Principal Defense Budget Formats, Fiscal Year 1985 13
2-5. The Defense Budget by Major Inputs, Fiscal Years 1981–84 14
2-6. The Fiscal 1986 Defense Budget by Major Mission 14
2-7. A Hypothetical Five-Year Defense Program, Fiscal Years 1984–88 17
3-1. National Defense Outlays, Fiscal Years 1929–85 19
3-2. U.S. Armed Forces Active-Duty Personnel, Selected Years, 1935–80 19
3-3. Fiscal Allocation between Nuclear and Conventional Forces, Selected Years, 1948–81 21
3-4. Defense Investments and Payrolls, Selected Fiscal Years, 1948–82 23
3-5. Cost of Carter's and Reagan's Five-Year Defense Programs in Current and 1986 Dollars, Fiscal Years 1982–86 24
3-6. Presidential Requests and Congressional Appropriations for the Defense Budget, Fiscal Years 1980–85 25
3-7. Outlays as a Percent of Gross National Product and Federal Spending, Fiscal Years 1977–85 26
3-8. Federal Transactions in the National Income and Product Accounts, Calendar Years 1977–85 26
3-9. Real Growth in the Accounts for Military Personnel, Operation and Maintenance, and Procurement, Fiscal Years 1980–85 27
3-10. Real Growth in Total Obligational Authority for the Strategic and General Purpose Forces, Fiscal Years 1980–85 27
3-11. Composition of National Defense Outlays, Fiscal Years 1980–85 28
3-12. Outlays as a Percent of Gross National Product and Federal Spending, Fiscal Years 1986–90 29
3-13. Real Growth in Investment and Operating and Support Accounts, Fiscal Years 1986–90 29
3-14. Proposed Budget Authority for Major Missions, Fiscal Years 1986–90 30
4-1. Periods of Real Growth in Defense Budget Authority and Outlays, Selected Fiscal Years, 1949–85 32
4-2. Examples of High-Priced Items 33
4-3. The Defense Budget in the Decade of Neglect, Fiscal Years 1971–81 38

4-4. Defense Investments by Five-Year Periods, Fiscal Years 1951–85 38
4-5. Acquisition of Major Weapons Systems, Fiscal Years 1974–85 39
4-6. Defense Personnel and Force Structure, Fiscal Years 1964, 1968, 1973 39
4-7. Defense Personnel and Force Structure, Fiscal Years 1976, 1980, 1985 42
4-8. Modernization of Military Equipment under Presidents Carter and Reagan, Fiscal Years 1977–85 43
4-9. Percentage of Equipment Capable of Performing Mission, Fiscal Years 1980, 1984 45
4-10. Military Training, Fiscal Years 1980, 1982, 1984 45
4-11. War Reserve Stocks of Conventional Munitions 46
4-12. Potential Soviet Deployment of General Purpose Forces by Theater of Military Operations 50
4-13. Selected Soviet Military Equipment Inventories and Production 53
4-14. Defense Intelligence Agency Count of NATO and Warsaw Pact Forces, 1983 54
5-1. Estimated Soviet Nuclear Capabilities, 1992 63
5-2. Number and Readiness of Warsaw Pact Divisions 64
5-3. U.S. and Allied Forces Available for the Defense of Central Europe 66
5-4. Salient Baseline, Programmed, and Combat Force Capabilities, Fiscal Year 1992 76
5-5. Cost of the Three Forces by Major Capability, Fiscal Year 1990 77
5-6. Cost of the Three Forces by Major Planning Contingency, Fiscal Year 1990 77
5-7. Five-Year Costs of the Three Forces (Retired Pay Excluded), Fiscal Years 1986–90 78
5-8. Summary of the Performance of the Three Forces 80
5-9. Performance of the Three Forces in Strategic Nuclear Retaliation, Day-to-Day Alert, 1981, 1982 82
5-10. Performance of the Three Forces in Tactical Nuclear Retaliation in Central Europe, Generated Alert, 1981, 1992 85
5-11. Performance of the Warsaw Pact and the Three Forces in the Defense of Central Europe, 1981, 1992 86
5-12. A U.S. Attack with Carrier Battle Groups on the Murmansk Area 87
5-13. Simultaneous Deployments of the Three Forces to Central Europe and the Persian Gulf, 1981, 1992 88
5-14. M + 90 Deployment of the Programmed and Combat Conventional Forces, Fiscal Year 1992 90
6-1. Summary of the Current Planning-Programming-Budgeting System 101
6-2. The Joint Chiefs of Staff's View of Minimum-Risk Forces, Fiscal Year 1991 101
6-3. Average Cost of a Carrier Battle Group 102

Figures

4-1. U.S. Defense Outlays and Estimated Dollar Cost of Soviet Defense Activities, 1951–81 36
4-2. Growth in Soviet Defense Program, 1970–81 37
4-3. U.S. and Soviet Defense Investments 49

CHAPTER ONE

DISARRAY IN DEFENSE

FOR ANYONE attempting to fathom the defense budget and its logic, 1985 was not a vintage year for rationality and lucidity. By the spring of 1985 most senators and representatives had decided to halt for at least one year what were to have been five more years of significant real growth in national defense budget authority, and after that to allow only a very gradual increase in resources.

Congress was not alone in considering the defense budget a target for restraint. As early as February 4, when the Reagan administration had submitted its overall federal budget, it calculated that a continuation of its current defense program in fiscal 1986 would require $324.8 billion in budget authority and $286.2 billion in outlays, while national defense would need $333.6 billion in budget authority and $294.6 billion in outlays. However, the president indicated that he would be able to get by with defense budget authority of $313.7 billion, national defense budget authority of $322.2 billion, defense outlays of $277.5 billion, and national defense outlays of $285.7 billion. Neither the president nor his secretary of defense, Caspar W. Weinberger, explained what justified these reductions or even what the difference was between national defense and defense.[1]

The Senate, brushing aside this bewildering array of figures, passed a budget resolution that specified budget authority of $302.5 billion and outlays of $273.1 billion for national defense, while the House of Representatives, in its own budget resolution, ordained $292.6 billion in budget authority and $267.1 billion in outlays for the same function. The two sets of numbers differed because the Senate added inflation to fiscal 1985 budget authority to arrive at its fiscal 1986 estimates, but the House

1. For an explanation of the difference, and of defense terminology in general, see chapter 2.

did not. These numbers also differed from the current services estimate and the presidential request for fiscal 1986 (table 1-1). Both House and Senate left to their authorization and appropriation committees the task of deciding where the administration's budget should be cut to comply with their resolutions.

To complicate matters further, the Defense Department must issue a five-year defense program, which projects its budget not just for the coming fiscal year but for the following four fiscal years. By contrast, the congressional budget resolutions propose three-year plans. The department estimated that defense budget authority would need to grow in real terms (that is, with inflation removed) by 5.9 percent in fiscal 1986, 8.2 percent in 1987, and 8.8 percent in 1988 in order to fund its program. The Senate and House, having allowed respectively zero and negative real growth in budget authority for national defense in fiscal 1986, agreed to 3 percent real growth in each of the next two fiscal years. But, of course, the base from which each measured the 3 percent increases was different. Consequently, as many different numbers existed for the so-called out-years as for fiscal 1986 (table 1-2). Reporters of the proliferation understandably became confused about which ones represented the defense budget.

To make matters worse, as spring turned to summer it became evident that however these differences might be resolved, the federal deficit had become an upwardly mobile target (table 1-3). The administration, in February, had projected real economic growth of 4 percent a year starting in 1985 and continuing through 1988. Based on that assumption and its own budget proposals, it had estimated that by the end of fiscal 1988 the federal deficit would fall from more than $200 billion and 5.7 percent of the gross national product to $144.4 billion and 2.9 percent of GNP. The House and Senate, using the same economic assumptions, set even more ambitious targets: the House voting for an estimated deficit in 1988 of no more than $124.4 billion; the Senate going even lower to $104.3 billion. Hardly had they done so, however, when the Office of Management and Budget announced that its original assumptions about economic growth, and hence about revenues and other factors affecting the size of the deficit, had been too optimistic. The upshot was that despite the herculean efforts of both House and Senate and all the cuts they had already made, the deficit would probably be almost as high in 1988 as it had been in 1985 and might still absorb a painfully large share of both GNP and personal savings.

Table 1-1. Defense Budget Estimates, Fiscal Year 1986
Billions of dollars

Item	*Current services*[a]	*Administration*	*Senate*[b]	*House*[b]
Budget authority				
Defense (051)	324.8	313.7	...	...
National defense (050)	333.6	322.2	302.5	292.6
Outlays				
Defense (051)	286.2	277.5	...	...
National defense (050)	294.6	285.7	273.1	267.1

Sources: *Special Analyses, Budget of the United States Government, Fiscal Year 1986*, pp. A-20, A-32; and *New York Times*, June 11, 1985.
a. The current services budget, in the case of defense, assumes that the administration's programs, as outlined in the previous fiscal year, would continue without change.
b. These are the budgets proposed by the budget committees of the Senate and the House of Representatives and passed by their respective bodies. The two committees officially propose budgets for national defense only.

Table 1-2. Five-Year and Three-Year Defense Plans, Fiscal Years 1986–90
Billions of dollars

Item	*1986*	*1987*	*1988*	*1989*	*1990*
Budget authority					
Administration					
Defense (051)	313.7	354.0	401.6	438.8	477.7
National defense (050)	322.2	363.3	411.5	448.9	488.1
Senate Budget Committee					
National defense (050)	302.5	323.4	346.8	...	...
House Budget Committee					
National defense (050)	292.6	312.7	335.4	...	...
Outlays					
Administration					
Defense (051)	277.5	312.3	348.6	382.3	418.3
National defense (050)	285.7	321.2	358.4	392.3	428.5
Senate Budget Committee					
National defense (050)	273.1	292.1	313.0	...	...
House Budget Committee					
National defense (050)	267.1	285.2	303.9	...	...

Sources: *Budget of the United States Government, Fiscal Year 1986*, pp. 9-13, 9-16; and *New York Times*, June 11, 1985.

Despite these attempts to cut the defense budget and the deficit, the main actors in the budgetary drama rarely suggested that the Defense Department did not need all the money it was requesting. Scandals in the defense industry as well as the increasingly obvious damage to the economy being wrought by the federal deficit help to explain why Congress decided to cut the administration's requests for national

Table 1-3. Deficit Estimates, Fiscal Years 1986–90
Billions of dollars in outlays

Agent and date of projection	*1986*	*1987*	*1988*	*1989*	*1990*
Current services (February 4, 1985)	230.3	245.6	247.8	232.8	224.4
Administration (February 4, 1985)	180.0	164.9	144.4	107.5	82.4
Senate Budget Committee (June 1985)	171.4	144.8	104.3	. . .	. . .
House Budget Committee (June 1985)	173.2	162.5	124.4	. . .	. . .
Office of Management and Budget (June 1985)[a]	188–91	. . .	175.0	. . .	. . .

Sources: *Special Analyses, Budget of the United States Government, Fiscal Year 1986*, p. A-14; and *New York Times*, June 11, 1985.
a. These estimates were based on the budget proposals of the Senate Budget Committee, but with more pessimistic assumptions about economic and federal revenues than had been used in the previous estimates.

defense. But no justification was given either for a freeze in budget authority for fiscal 1986, or for limiting real growth to 3 percent a year after that. Nor did Congress attempt to justify the exclusion of Defense from still further cuts, despite the intractable deficit, and despite the revelation that prior-year programs had been substantially overfunded and the Pentagon had money to spare for future investment.

President Reagan did not clarify matters by his role in the process. As the congressional knives began to flash, he announced that the 5.9 percent real increase in defense he had requested for fiscal 1986 was the minimum consistent with the maintenance of U.S. security. After congressional surgery had begun, he admitted that the Pentagon could somehow manage with a 3 percent real increase. Once the congressional resolutions had passed, he discovered that the fiscal 1985 defense budget, with an allowance for inflation, would satisfy the country's needs in fiscal 1986.

Perhaps the political process makes this lack of intelligible discourse inevitable. Perhaps defense, like Winston Churchill's Russia, is bound to remain "a riddle wrapped in a mystery inside an enigma" for those inside as well as outside the process. But it would be tragic to treat that possibility as a certainty. While the language of the defense budget may sound arcane, its logic is easy enough to understand. There is no mystery about the trends in defense spending that brought the country physically unscathed through nearly forty years of the nuclear age. The magnitude of the defense budget buildup under the Reagan administration during the last four years and the reasons for the reaction to it are not difficult to grasp. The ability to compare the efficiency and sufficiency of various defense budgets is not entirely out of reach. Even an understanding of why efficient solutions to military problems might be rejected, and what might be done about it, is possible for taxpayers—in or out of uniform—to obtain.

In short, defense need not be nonsense. Winston Churchill, always quotable, recalled in his memoir of World War I and its antecedents that when he was First Lord of the Admiralty he had once asked for six new dreadnoughts. David Lloyd George, then Chancellor of the Exchequer, insisted that no matter what the Germans were building, Great Britain could afford no more than four. After listening to this debate the Asquith cabinet finally compromised on eight. Attempts to improve on that kind of performance may prove difficult in the current climate. Still, there is much to be said for looking at the possibilities.

CHAPTER TWO

CONTROLLING THE DEFENSE BUDGET

AN UNDERSTANDING of the language and development of the defense budget is essential to an appreciation of what can be done to control the process. However, what constitutes the cost of defense is not immediately self-evident. The federal budget, for example, contains a number of different accounts that sound as though they had something to do with national security (table 2-1). Which accounts should be included in a definition and analysis of defense functions is frequently a source of debate and confusion.

Definitions of Defense

The military functions of the Defense Department, which constitute what is normally defined as the defense budget, are carried in the 051 account. They will entail outlays of $246.3 billion in fiscal 1985. Because of historical circumstances, most notably the establishment of an independent Atomic Energy Commission (now part of the Department of Energy), several other functions with a direct bearing on defense are located in other agencies. These functions consist of the military applications of atomic energy, civil defense, the stockpiling of strategic materials, and the selective service system. The federal budget in its functional format (as opposed to its organization by agency) combines these activities with the military functions of the Defense Department in an account labeled national defense and numbered 050. Because the 051 and 050 accounts have roughly similar titles, and because the budget committees of the House and Senate deal only with the 050 account in their annual resolutions, the budget for national defense is frequently confused with the budget for defense (military functions). However, the

Table 2-1. Budget Authority and Outlays by Function, Fiscal Years 1984–86
Billions of dollars

Item	*1984*	*1985*	*1986*
Budget authority			
National defense (050)	265.2	292.6	322.2
Department of Defense (military functions)	258.2	284.7	313.7
Other	7.0	7.8	8.5
Department of Defense (civil functions)	19.2	30.0	32.6
Veterans' benefits	26.5	27.3	27.4
Outlays			
National defense (050)	227.4	253.8	285.7
Department of Defense (military functions)	220.8	246.3	277.5
Other	6.6	7.5	8.2
Department of Defense (civil functions)	19.5	19.0	20.3
Veterans' benefits	25.6	26.9	26.8

Source: *Budget of the United States Government, Fiscal Year 1986*, pp. 9-12, 9-13, 9-15, 9-16.

national defense account (050) is currently larger than the defense account (051) by roughly $8 billion.

Confusion about the costs of defense can also arise because of the civil functions performed by the Defense Department. These functions are shown in accounts that are separate from the national defense and defense functions. They consist primarily of the nonmilitary activities of the Corps of Engineers, cemeterial expenses, and the transactions of the military retirement system. All together they will entail expenditures of around $20 billion in fiscal 1985.

It is also possible to argue that any review of national security needs and costs should take account of military assistance, benefits and services for veterans, and the payment of interest on that portion of the public debt caused by past wars. All but the last are shown as separate accounts in the federal budget. Military assistance (now described as international security assistance) amounts to roughly $8 billion in outlays and is appropriated to the president. Most of the funding for veterans goes directly to the Veterans' Administration and entails annual expenditures of about $27 billion. Estimates can be found of the costs of past U.S. wars, including current interest on loans incurred during earlier wars, but they are not especially reliable.

All or some of these functions can be defined as related to defense. However, if the main purpose of an analysis is to understand current national security needs and costs, and to let bygones be bygones, focusing primarily on the defense (051) and national defense (050) accounts makes sense. In principle, military assistance should also be included, as it used to be when it was part of the Defense Department's budget. It would be difficult to argue, however, that military assistance has been used as a serious substitute for U.S. military power since the late 1960s. For fifteen years or more its allocation has been related more to winning friends and influencing people than to defense.

It is as important to understand how the various defense accounts are funded, and the funds spent, as it is to define the accounts under consideration. The Defense Department has the largest payroll in the federal government. It spends more on the operation and maintenance of its equipment than any other agency. It is also the government's biggest investor: in construction, research and development, and the acquisition of millions of products ranging from intercontinental ballistic missiles to toothpaste. For the most part there is no market in which suppliers of weapons, equipment, and munitions haggle with defense buyers over prices and quantities of these items. Much of what the department buys is not in stock; it must be custom-made, usually to detailed defense specifications, takes years (eight for a nuclear-powered aircraft carrier) to build, and is expensive.

Budget Authority

Because armaments take so long to manufacture and cost so much, the Defense Department must be able to assure its contractors that it has the funds to pay for their work, make payments for work in progress, and frequently minimize the capital investments of manufacturers by providing them with government-furnished equipment. These needs lead to the particular importance of distinguishing between budget authority and outlays when looking at defense. Budget authority performs two functions. It gives the Defense Department the authority to sign contracts for activities authorized by Congress, and it tells the Treasury that it must be prepared to pay bills up to the amounts of budget authority appropriated by Congress.

Once this budget authority becomes available, Defense is able (subject to certain controls set by the Office of Management and Budget) to negotiate explicit contracts, hire or retain personnel, and draw on the Treasury to pay its bills as they are presented. These payments, which may occur years after the appropriation of budget authority for a particular program, constitute outlays or expenditures and have a major impact on both the economy and the size of the federal deficit.[1]

As the economic influence of federal outlays has grown, Congress has become increasingly concerned about controlling them. Indeed, it now sets expenditure targets in its annual budget resolutions. But Congress only authorizes and appropriates budget authority to the Defense Department and other agencies; it does not appropriate outlays. What is more, as items have grown more costly, as the time required for their production has lengthened, and as long-term planning has been forced on defense— if for no other reason than to show the future costs of current programs—Congress has taken increasingly to what is known as the full funding of items in these programs. Thus, when the Navy set a goal of a 600-ship fleet and requested 2 new nuclear-powered aircraft carriers, Congress agreed; it fully funded both carriers in fiscal 1983 and provided budget authority of $6.8 billion for their construction. Final payments from that authority probably will not be made until 1990 or 1991.

Congress (which can do almost anything) can also agree to the multiyear procurement of such items as tanks, helicopters, and aircraft, which the Defense Department will want to buy in large quantities over a number of years. The Air Force, for example, can go to an aircraft manufacturer and contract for the production of 100 fighters a year for the next five years, with both parties to the contract secure that Congress will provide the necessary budget authority from year to year. This procedure is meant to result in lower prices because the producer is assured of longer, more orderly, and more efficient production schedules. Even so, congressional acceptance of multiyear procurement has been decidedly more gradual than the Defense Department had hoped. Not only have members of both the House and Senate been loath to surrender their authority to make annual budget decisions; they have also questioned whether the advantages of multiyear procurement outweigh the loss of flexibility in adjusting the defense budget to economic and other changes. And representatives and senators continue to wonder whether these contracts are not more a device to commit Congress to controversial defense programs than a means of saving money.

Students of the defense budget distinguish between "fast" and "slow" money. As much as 60 percent of the current defense budget consists of

1. Of course, many spending decisions are also affected as budget authority is obligated in the form of contracts and corporate order books begin to fill up.

fast money, which is budget authority that results in a high rate of expenditure during the year in which it is appropriated. Examples of fast money are the payroll for military and civilian personnel, operating and maintenance expenses, and a great deal of research and development programs that are usually funded on a year-to-year basis. The slow money is concentrated in major procurement and military construction, programs that frequently are fully funded in a given year and hence result in outlays over an extended period. Table 2-2 draws the distinction between fast and slow money and estimates the rate at which budget authority, in many categories, converts to outlays.

Balances of Budget Authority

Most of this budget authority will expire automatically if not obligated in the form of explicit or (for some payrolls) implicit contracts within a fixed period. However, unspent authority will accumulate in Treasury balances, and, unless it expires or is rescinded by Congress, will continue to spend out in subsequent years regardless of decisions made about annual budgets (table 2-3). Consequently, defense outlays are always more difficult to control than budget authority; outlays coming from the balances of prior-year budget authority are usually classified as uncontrollable, although that need not always be true; and total outlays in any given year are almost always likely to differ from budget authority. When budget authority is increasing, as it has been recently, outlays will be lower than budget authority during the years of the increases. When budget authority is declining, as has frequently been true during the final years of a war and in its immediate aftermath, outlays may exceed budget authority in one or more years. However, it is not legally possible for outlays cumulatively to exceed the amount of budget authority made available from current and prior-year appropriations.

It may be legal but it is not realistic to expect, as some members of Congress do, that the unobligated balances of prior-year budget authority can be readily or fully recovered or used to keep current-year appropriations lower than might otherwise have been the case. All budget authority is designated by Congress for funding particular programs. Some portion of it may remain unobligated because a program has gone sour. Another portion may await obligation pending the negotiation of contracts. Increasingly, it could be that as inflation has declined, more

Table 2-2. Outlays, by Appropriation Categories and Major Weapons, as a Percent of First-Year Budget Authority Spent Each Year over a Six-Year Period

	Year					
Category of expenditure	*First*	*Second*	*Third*	*Fourth*	*Fifth*	*Sixth*
"Fast money"						
Personnel	97.7	1.9	0.4	...	...	...
Operation and maintenance	83.1	14.0	1.9	1.0	...	...
Family housing	59.8	25.0	8.8	4.0	2.4	...
Research, development, test, and evaluation	56.4	35.8	5.6	2.2	...	...
"Slow money"						
Procurement						
Missiles	19.1	49.1	27.6	4.2	...	...
Ammunition	12.1	52.2	28.4	7.3	...	...
Aircraft	10.1	44.3	35.9	6.0	3.7	...
Tanks	6.3	39.8	38.3	11.0	4.6	...
Ships	2.0	14.0	18.0	18.0	18.0	18.0
Military construction	7.7	36.4	30.4	12.7	6.5	6.3

Source: William W. Kaufmann, "The Defense Budget," in Joseph A. Pechman, ed., *Setting National Priorities: The 1983 Budget* (Brookings, 1982), p. 60.

Table 2-3. Balances of Prior-Year Budget Authority for Defense, End of Fiscal Years 1980–86

Billions of dollars

Budget authority	*1980*	*1981*	*1982*	*1983*	*1984*	*1985*	*1986*
Obligated	67.9	86.3	107.6	128.7	153.5	191.9	224.0
Unobligated	24.2	26.5	34.6	43.4	51.6	51.5	55.7
Total	92.1	112.8	142.2	172.1	205.1	243.4	279.7

Sources: *Budget of the United States Government, Fiscal Year 1982*, p. 557; *Fiscal Year 1983*, p. 9-9; *Fiscal Year 1984*, p. 9-11; *Fiscal Year 1985*, p. 9-16; and *Fiscal Year 1986*, p. 9-21.

budget authority was requested than has turned out to be needed for the completion of some programs. But whatever the case, the threat to rescind or reprogram prior-year balances could cause them to be obligated in an inefficient manner. Hidden gold may exist in these balances; it is not always easy to extract.

Budget Formats

Budgets, of course, are intended to do more than provide highly aggregated statements of what will be needed in budget authority and what outlays will occur in a given fiscal year. They are also meant to assist policymakers in the allocation and control of resources. Obviously, putting resources into particular categories will not by itself resolve the issues of allocation, but it will inform and guide planners in their thinking about the issues.

What categories to use has occasioned fierce debate in the past. However, there is no single right way to classify the defense budget. What is right depends on what issues and choices the policymakers wish to emphasize. In recognition of this simple truth, the Defense Department produces the same budget totals organized and allocated in three different ways (table 2-4). The most familiar is the budget by appropriation title mandated by Congress. It organizes appropriations of budget authority according to what the Defense Department buys. These purchases of goods and services are classified under such unwieldy titles as military personnel, operation and maintenance, procurement, and research, development, test, and evaluation. It would be difficult to demonstrate that the information in this budget helps policymakers to make major defense decisions, although it tells a great deal about how and where defense funds will be committed.

The same appropriations are contained in a second budget organized according to such major components of the Defense Department as the Army, Navy, and Air Force. If nothing else, this format shows what share of the budget each of the services receives and whether the distribution is changing over time. If shares in the three principal "corporations" were to be sold, the Navy would probably be seen as the Ma Bell of old because of its diversified holdings and relative stability; the Air Force as the epitome of the high-tech company; and the Army as the most cyclical of the three—doing well in war but not a good investment in peacetime.

A program budget, first introduced in 1962, allocates appropriations to such major capabilities and missions as the strategic nuclear forces, the general purpose (mostly conventional) forces, intelligence and communications, airlift and sealift, research and development, central supply and maintenance, and administration and associated activities. The program budget was intended to force decisions about these capabilities and missions independently of the service that controlled them. Whether the result has ever matched the intention continues to be in dispute.

Allocations can be organized in other, still more interesting categories than those used by the program budget (table 2-5). However, the Defense Department frowns on such exercises, largely because they are thought

Table 2-4. Principal Defense Budget Formats, Fiscal Year 1985
Billions of dollars

Item	*Budget authority*
Appropriation title	
Military personnel[a]	68.9
Operation and maintenance	78.2
Procurement	96.8
Research, development, test, and evaluation	31.5
Military construction	5.5
Military family housing	2.9
Revolving and management funds	1.7
Receipts and deductions	−0.6
Total	284.7
Component	
Department of the Army	74.4
Department of the Navy	96.5
Department of the Air Force	99.9
Defense agencies, OSD, JCS[b]	13.0
Defense-wide	1.0
Total	284.7
Program	
Strategic forces	27.8
General purpose forces	120.6
Intelligence and communications	25.1
Airlift and sealift	7.0
National Guard and Reserve	15.7
Research and development	24.6
Central supply and maintenance	24.4
Training, medical, and other general personnel activities	33.1
Administration and associated activities	5.9
Support of other nations	0.5
Total	284.7

Sources: *Budget of the United States Government, Fiscal Year 1986*, pp. 5-5, 5-6; and *Department of Defense Annual Report to the Congress, Fiscal Year 1986*, p. 294. Numbers are rounded.
a. This category now includes funds for the military retired pay accrual account.
b. OSD stands for Office of the Secretary of Defense; JCS stands for Joint Chiefs of Staff.

to invite intrusion by Congress into decisions seen as the prerogative of the executive branch. Although it is quite feasible to show the major nuclear and conventional contingencies used by the Defense Department for planning purposes, the forces related to those contingencies, and their cost, officials fear that Congress might not only reject some of these planning contingencies but also delete the forces associated with them and dictate a different strategy (budget authority for the planning contingencies is estimated in table 2-6).

Table 2-5. The Defense Budget by Major Inputs, Fiscal Years 1981–84
Billions of 1985 dollars unless otherwise noted

Item	*1981*	*1982*	*1983*	*1984*	*Percent real growth*
Personnel and retired pay	68.3	71.0	72.7	75.1	10.0
New weapons and equipment	57.6	70.4	85.7	89.1	54.7
Military construction	27.7	29.9	30.7	31.7	14.4
Maintenance and spare parts	18.3	20.6	22.9	26.2	43.2
Transport and logistics	17.3	19.3	18.8	18.8	8.7
Operations and training	15.8	17.6	18.3	19.0	20.3
War reserve stockpile	8.3	11.3	10.4	10.8	30.1

Source: Richard Halloran, "U.S. Forces May Lack Resources for Sustained War," *New York Times*, May 14, 1984.

Table 2-6. The Fiscal 1986 Defense Budget by Major Mission[a]
Billions of dollars

Mission	*Budget authority*[b]
Strategic nuclear retaliation	51.5
Theater nuclear retaliation	3.2
Conventional defense of:	
Central Europe	80.2
North Norway	17.2
Greece and Turkey	9.8
Atlantic and Caribbean[c]	25.8
Persian Gulf states	20.9
Republic of Korea	12.9
Pacific and Indian Oceans	21.7
Continental United States, Alaska, and Panama[c]	16.2
Intelligence and communications	36.5
Subtotal	295.9
Retired pay accrual	17.8
Total budget authority	313.7

Sources: *Budget of the United States Government, Fiscal Year 1986*, p. 5-6; and author's estimates.
a. The missions and their costs reflect the main contingencies for which the United States plans and the forces associated with those planning contingencies. Reality and the planning contingencies may never coincide.
b. Totals include indirect as well as direct costs of the forces. Consequently, they are higher than the totals shown in the program budget.
c. These totals are so high because they include the costs of forces unable to deploy overseas in a timely fashion or undergoing maintenance and training.

The Budget Cycle

The preparation, passage, and implementation of defense budgets is an industry in itself and has become a continuous process. Preparation of a particular budget begins a year before the document is ready for submission. The president presents the request to Congress by the end of January or the beginning of February. During the next eight months the congressional budget committees must review it before passage of a budget resolution, the authorization committees must legislate the boundaries within which appropriations are to be made, and the appropriation committees must decide on the amounts of budget authority to make available within these limits. The fiscal year begins on October 1 and ends on September 30 of the following calendar year; it is numbered according to the year in which it ends, to add to the confusion. Hence fiscal 1985 began on October 1, 1984, and ended on September 30, 1985. If Congress has not completed action on the budget by October 1, it must pass a continuing resolution allowing the department to operate until the new appropriations become available.

Nearly three years will have elapsed from the beginning of this cycle to the end of the fiscal year. Inevitably, many assumptions made in the budget will have proved wrong, because of congressional action, domestic circumstances, or external conditions. Consequently, both the Defense Department and Congress must be able to change both the size of the budget and its allocation of resources. Additional funds can be acquired by the passage of amendments to a budget under consideration by Congress or through supplemental appropriations at any time during the fiscal year, although Congress now prefers to consider such requests at the same time that it acts on the budget for the next fiscal year. The department, with the approval of the four main committees having defense responsibilities, may also reprogram resources from one line item or account to another, and either the department or Congress (on its own initiative) may move to rescind previously appropriated budget authority, including prior-year authority whether obligated or not.

If rescissions are not made, unused budget authority may, in principle, automatically lapse.[2] In practice, the authority is transferred to an "M" account where it remains available for the payment of residual obligations and liabilities chargeable to various appropriation accounts. That is what has happened to some of the unused budget authority accumulated during the Reagan years. Presumably it, too, can be reprogrammed with congressional consent.

2. For one-year accounts, the unobligated balance expires at the end of the fiscal year. For multiple-year accounts the unobligated balance may be carried forward and remain available for the period specified. In no-year accounts, the unobligated balance can be carried forward indefinitely unless rescinded or unless disbursements have not been made against the appropriations for two full years.

The Five-Year Defense Program

For orderly long-range planning, and to ensure that near-term commitments do not exceed the capacity of later budgets to meet them, the Defense Department prepares a five-year defense program (known as the FYDP) with its annual budget. This rolling FYDP is an extension of the program budget. It details what the department plans to request during the five-year period and how it proposes to allocate its resources among the various investment and operating accounts. In effect, it portrays how U.S. forces will evolve and what their costs will be. A hypothetical FYDP is shown in table 2-7.

The FYDP does not constitute a formal commitment by Defense to these totals and allocations, and it is revised each year as an old fiscal year is dropped from the past and a new one added to the future. But despite its provisional nature, the program is meant to force policymakers to think seriously and systematically about the future. If conscientiously implemented, the FYDP minimizes the probability that "foot-in-the-door" tactics can be used to shoehorn new and potentially budget-busting weapons into the budget.

Most weapons programs begin with what, by defense standards, are modest demands for resources, and a large number of them can initially be accommodated without much change in funding patterns. However, each of these programs has a propensity to balloon rapidly in costs even before it reaches the expensive stage of procurement. Indeed, it is no longer unusual to spend billions of dollars on the research, development, test, and evaluation of a major weapon system. If a secretary of defense does not take account of this propensity, and of the pervasive tendency to underestimate the costs of new capabilities, he will quickly find himself with more new programs on his hands than he can reasonably expect to fund (or fund at efficient rates) in later years. Unless the secretary controls this process at an early stage, the consequences are likely to be the widespread cancellation of programs, their costly stretch-out, or the even more costly practice of stopping and starting them. The FYDP, if properly used, can avoid such consequences by conservatively spelling out the future costs of ongoing and planned investments and by clarifying the choices that will have to be made within inevitable budgetary constraints.

Such choices are always unpleasant and difficult, since most new

Table 2-7. A Hypothetical Five-Year Defense Program, Fiscal Years 1984–88[a]
Billions of dollars of budget authority

Program	*1984*	*1985*	*1986*	*1987*	*1988*
Strategic forces	26.1	27.8	29.9	30.5	32.1
General purpose forces	100.7	120.6	132.1	152.5	177.4
Intelligence and communications	20.0	25.1	27.9	31.2	33.9
Airlift and sealift	5.5	7.0	8.0	8.5	8.0
National Guard and Reserve	12.2	15.7	16.9	19.9	22.0
Research and development	21.5	24.6	30.4	33.6	40.6
Central supply and maintenance	23.4	24.4	26.5	30.6	35.5
Training, medical, and other general personnel activities	43.3	33.1[b]	35.6	39.7	44.0
Administration and associated activities	4.8	5.9	5.9	6.6	7.2
Support of other nations	0.7	0.5	0.5	0.9	0.9
Total budget authority	258.2	284.7	313.7	354.0	401.6

Source: *Budget of the United States Government, Fiscal Year 1986*, p. 5-6.

a. The Defense Department does not provide an unclassified version of its five-year defense program. Consequently, this hypothetical example consists of the amounts appropriated by Congress for fiscal 1984 and 1985 and the amounts proposed for the coming three fiscal years (1986, 1987, 1988), as published in the fiscal 1986 budget.

b. Before fiscal 1985, military retired pay was included in this category. Starting in fiscal 1985, with adoption of the accrual system, retired pay has been distributed to the mission categories.

weapons systems would be nice to have. But a secretary of defense who is unable to distinguish between a luxury and a necessity is unlikely to end up with a reasonable defense.

CHAPTER THREE

TRENDS IN NATIONAL DEFENSE

IT IS HARDLY surprising that defense planning and budgeting should have undergone a considerable transformation during the last fifty years. During those same years the size of the defense budget was, in a series of major steps, undergoing a major metamorphosis. The growth of national defense outlays during the last fifty-seven years illustrates the magnitude of the metamorphosis (table 3-1). Significant change also occurred in active-duty military personnel for selected fiscal years during this same period (table 3-2).

Isolation

U.S. outlays for national defense, measured in fiscal 1986 dollars, hovered between $6.5 and $11.6 billion from 1929 through 1938 and did not rise above the $12 billion mark until 1939 and the outbreak of war in Europe. Costs remained so low because the armed forces consisted of roughly 250,000 active-duty personnel and because the Army and the Navy were allowed to buy relatively little in the way of weapons and equipment. Such parsimony was deemed desirable and feasible because a majority of Americans still believed that the United States could isolate itself from foreign quarrels and because these quarrels, for the most part, were geographically distant. The shield provided by France and Britain, combined with distance, meant that the armed forces would have the time to mobilize and expand in the unlikely event that a threat to the Western Hemisphere began to develop.

World War II and technology changed all that. The United States became not just the arsenal of democracy but also its principal defender. Long-range aircraft, the possibility of intercontinental missiles, and nuclear weapons simultaneously shrank distance and reduced the time

Table 3-1. National Defense Outlays, Fiscal Years 1929–85
Billions of 1986 dollars

Year	Outlays	Year	Outlays	Year	Outlays
1929	8.8	1948	67.0	1967	255.4
1930	9.8	1949	92.5	1968	278.8
1931	9.8	1950	91.2	1969	270.9
1932	10.2	1951	143.6	1970	249.6
1933	8.8	1952	258.1	1971	226.3
1934	6.5	1953	285.0	1972	209.8
1935	8.8	1954	271.3	1973	190.5
1936	10.7	1955	231.5	1974	181.9
1937	10.7	1956	219.3	1975	180.8
1938	11.6	1957	222.8	1976	175.1
1939	12.1	1958	217.8	1977	177.8
1940	19.1	1959	218.8	1978	178.6
1941	63.2	1960	213.6	1979	185.2
1942	213.9	1961	214.3	1980	190.4
1943	512.3	1962	225.3	1981	199.5
1944	625.1	1963	228.5	1982	216.0
1945	692.5	1964	225.4	1983	234.1
1946	358.1	1965	203.9	1984	245.4
1947	97.9	1966	219.8	1985	264.1

Sources: U.S. Bureau of the Census, *Historical Statistics of the United States, Colonial Times to 1970*, pts. 1 and 2 (Government Printing Office, 1975), pp. 230, 1115-16; *Historical Tables, Budget of the United States Government, Fiscal Year 1986*, pp. 6.1(1)–6.1(8); and author's estimates.

Table 3-2. U.S. Armed Forces Active-Duty Personnel, Selected Fiscal Years, 1935–80
Thousands

Year	Number	Year	Number
1935	252	1964	2,687
1945	12,123	1968	3,548
1948	1,446	1974	2,162
1956	2,806	1980	2,063

Sources: Department of Defense, Washington Headquarters Services, Directorate for Information, Operations, and Reports, *Selected Manpower Statistics, FY 1979*, p. 60; and Department of Defense, Office of the Assistant Secretary of Defense (Comptroller), *National Defense Budget Estimates for FY 1986*, pp. 116–17.

available for mobilization. The postwar debility and dependence of Western Europe meant that no great powers were in place to hold the front lines while U.S. forces prepared for war.

The Truman Years

World War II had resulted in peak outlays for national defense of $692.5 billion (fiscal 1986 dollars) and armed forces of more than 12 million men and women. By fiscal 1948, President Harry S. Truman had brought active-duty military personnel down to fewer than 1.5 million and was attempting to stabilize national defense outlays in the neighborhood of $70 billion (fiscal 1986 dollars). That this total was a factor of seven higher than before the war resulted from what were seen as the changed conditions of U.S. security. (The American government was becoming a party to a growing number of mutual defense treaties and was wary of the threat to them from the Soviet Union and China.) That the amount was not even greater was attributed primarily to the U.S. monopoly of atomic weapons, a monopoly that led the first secretary of defense, James V. Forrestal, to write:

> As long as we can outproduce the world, can control the sea and can strike inland with the atomic bomb, we can assume certain risks otherwise unacceptable in an effort to restore world trade, to restore the balance of power—military power—and to eliminate some of the conditions which breed war. The years before any possible power can achieve the capability effectively to attack us with weapons of mass destruction are our years of opportunity.[1]

From Eisenhower to Reagan

Forrestal wrote those words in 1947. Only two years later the Soviet Union detonated an atomic device; by the middle of 1950 the United States was engaged in a bloody but limited war in Korea. This sequence of events led to another upward step in national defense outlays.

The Korean War had entailed peak expenditures of $285 billion (fiscal 1986 dollars). The Eisenhower administration was able to bring the total down to $219 billion by fiscal 1956 and to stabilize active-duty military personnel at about 2.8 million. Even so, the new constant dollar level was three times higher than the one that had existed so briefly before the war. The president and his secretary of state extolled the "New Look" they had taken at defense plans and programs; they insisted that nuclear explosives should now be considered conventional weapons; they uttered threats of massive retaliation in the event of Soviet or Chinese aggression. But it was the need to maintain large conventional as well as nuclear forces that accounted for so much of the real increase. Admittedly, total obligational authority for the strategic nuclear forces rose from 12.6 percent of the defense budget in fiscal 1948 to 23.9 percent in

1. Walter Millis, ed., *The Forrestal Diaries* (Viking Press, 1951), pp. 350–51.

Table 3-3. Fiscal Allocation between Nuclear and Conventional Forces, Selected Fiscal Years, 1948–81
Billions of 1986 dollars

	Total obligational authority	
Year	*Nuclear forces*[a]	*General purpose forces*[b]
1948	9.4	38.2
1952	60.4	151.1
1956	46.5	80.9
1961	53.2	77.1
1971	20.6	82.8
1979	13.0	87.7
1981	15.3	102.8

Source: Department of Defense, *National Defense Budget Estimates for FY 1986*, p. 69.
a. Consists of program I (strategic forces) funds only.
b. Consists of program II (general purpose forces), program IV (airlift and sealift), and program V (national guard and reserve forces) funds.

fiscal 1956 (as measured by the allocation rules of the program budget and shown in table 3-3).[2] But while the share of the budget allocated to the general purpose forces fell from 45.3 percent to 41.6 percent in the same years, total obligational authority for these forces increased in real terms by a factor of more than two. Other programs such as intelligence and communications, and research and development also increased in real terms as well as in percentage of the total budget.

The allocation between nuclear and conventional forces would change again—and in favor of the conventional forces—with the passage of time. Equally important, the $219 billion in outlays for national defense in fiscal 1956 was to come close to the average for the next twenty-five years. During the three years of the Kennedy administration real outlays for national defense increased by 7 percent, and at the high point of the war in Southeast Asia (fiscal 1968) they were 28 percent higher than a decade earlier. But by fiscal 1981, after a rapid postwar decline, outlays were back at $200 billion and rising toward the twenty-five-year average of $213 billion. Meanwhile, of course, defense spending declined as a percent of a growing GNP, thus making it less of a burden on the nation.

During those same twenty-five years total obligational authority for

2. Total obligational authority is a term unique to the Department of Defense; TOA is a financial measurement that may include some prior-year funding and other financial adjustments. It may be somewhat larger or smaller than budget authority.

the strategic nuclear forces, as defined by the program budget, fell from 26.8 percent of the defense budget to 7.1 percent, while the share of the general purpose forces and related programs (airlift and sealift and guard and reserve forces) climbed from 40.5 percent to 46.3 percent. There were other changes as well. Active-duty military personnel declined between 1956 and 1981 from 2.8 to 2.1 million, largely because of the end of conscription. At the same time the defense payroll (for civilian personnel, active, reserve, and retired military), which was 52 percent of the total defense budget in fiscal 1956, fell to 42 percent in fiscal 1981, even though the total payroll in constant dollars changed very little—from $103.1 to $93.9 billion (table 3-4). Investment, on the other hand, rose in real terms from $53.1 billion to nearly $89 billion, an increase of 67 percent.

It would be difficult to argue that outlays for national defense averaging $213 billion over so many years were exactly the right prescription for U.S. security in a relatively unstable world. In retrospect, however, the prescription seems not to have damaged the patient. Those who look back on the 1950s and early 1960s as a golden age highlighted by American military superiority, since thrown away, would be hard put to show how, even with much larger defense budgets, the nuclear stalemate so evident by 1981 could have been averted. They would find it equally difficult to explain why the Soviet Union was so rambunctious in Berlin and Cuba during the golden age yet relatively so cautious during the 1970s, the alleged decade of neglect.

Part of the explanation could be that American superiority during the golden age was much less meaningful than it has since been cracked up to be. Even during the decade of neglect, U.S. and allied conventional forces could still give an impressive account of themselves in areas of vital interest. That this probably was true—that the odds did not shift radically in favor of Soviet aggression—should not be that surprising. For during the golden age, Nikita S. Khrushchev had performed such radical surgery on the armed forces of the USSR that it probably took more than a decade of real budget increases to compensate for his cuts. By the end of the 1970s, however, it was also true (in all probability) that Leonid I. Brezhnev had restored Soviet military power to the point where—if Russian budgets continued to expand and the U.S. defense effort continued to remain relatively flat—the odds of maintaining the traditional conditions of U.S. security were almost certain to decline. Indeed, Harold Brown, secretary of defense in the Carter administration,

Table 3-4. Defense Investments and Payrolls, Selected Fiscal Years, 1948–82
Billions of 1986 dollars

	Budget authority	
Year	*Investment*[a]	*Payroll*[b]
1948	18.4	57.4
1956	53.1	103.1
1965	73.7	97.1
1975	54.4	93.5
1980	71.0	92.7
1981	88.6	93.9
1982	108.6	96.7

Source: Department of Defense, *National Defense Budget Estimates for FY 1986*, pp. 87–89.
a. Consists of procurement; research, development, test, and evaluation; and military construction.
b. Consists of military and civilian pay, and retired military pay.

argued at the beginning of 1979 "that the gap between U.S. and Soviet defense expenditures cannot continue to expand without a dangerous tilt in the relevant balances of power and a weakening of the overall U.S. deterrent."[3] To forestall the danger he proposed real growth of 3 percent a year in defense budget authority for at least as long as the Kremlin continued to increase Soviet military resources.

The Reagan Revolution

When President Reagan entered office in 1981, he demanded even more dramatic changes, largely on the ground that the "tilt" feared by Harold Brown had already occurred some years earlier. Although the new administration was able to add $5 billion in total obligational authority to the defense budget for fiscal 1981 (the fiscal year already under way), the budget for fiscal 1982 was the first one on which it could have a major impact. President Carter, in his last defense recommendations, had proposed $196.4 billion in total obligational authority for fiscal 1982 and a FYDP total of $1,276.1 billion. After a few months in office, President Reagan requested that the total for fiscal 1982 be increased by 13.1 percent to $222.2 billion and that the FYDP total be raised by 14.4 percent to $1,460.2 billion (table 3-5).

This original plan underwent some modification in successive FYDPs. Indeed, projected total obligational authority for fiscal 1985 shrank from

3. *Department of Defense Annual Report, Fiscal Year 1980*, p. 17.

Table 3-5. Cost of Carter's and Reagan's Five-Year Defense Programs in Current and 1986 Dollars, Fiscal Years 1982–86

Billions of dollars

	Total obligational authority					*Five-year total*
Item	*1982*	*1983*	*1984*	*1985*	*1986*	
Current dollars						
Carter	196.4	224.0	253.1	284.3	318.3	1,276.1
Reagan	222.2	254.8	289.2	326.5	367.5	1,460.2
1986 dollars						
Carter[a]	228.9	240.4	252.4	260.1	278.2	1,260.0
Reagan	247.8	284.1	312.1	339.6	367.5	1,551.1

Sources: *Department of Defense Annual Report, Fiscal Year 1982,* p. 10; Alice Maroni and Robert Foelber, *The Defense Spending Debate: Comparing Recent Defense Appropriations with 1981 Projections,* Congressional Research Service, Library of Congress, Report 84-97F (May 29, 1984), p. 13; and author's estimates.

a. These constant-dollar calculations remove the inflation originally anticipated by the Carter administration and convert those results to 1986 dollars, using the same deflators as were estimated for the Reagan five-year defense program.

$326.5 billion to $305.7 billion between 1981 and 1984. But a great deal of this apparent self-restraint resulted from a substantial reduction in inflation and the strong dollar. Nevertheless, despite the reductions in the original requests, the administration's proposals represented a real increase in total obligational authority during its first four years (fiscal 1982 through fiscal 1985) of 43.1 percent, or an average annual real increase of 9.4 percent. By contrast, real growth in total obligational authority for defense during the previous four years (fiscal 1978 through fiscal 1981) was 12.2 percent, or an average annual real increase of 2.9 percent, most of which resulted from congressional additions to the Carter administration's last two budgets.

The Reagan administration ended up requesting $1,060.0 billion in total obligational authority for defense during its first four years. Congress reduced this total by 6.4 percent, or $67.4 billion (table 3-6). The bulk of the reduction—about 70 percent—came from stretching out some procurement programs and deferring others. Only 6 percent of the cuts resulted from outright cancellations. Most of the remaining savings grew out of lower inflation, reduced prices, and smaller payrolls than the administration had requested.

Despite these adjustments, real growth remained impressive. In four years defense budget authority increased by 35.0 percent, or close to 8 percent a year on the average, while defense outlays rose by 31.5 percent, or more than 7 percent on the average. By the end of fiscal 1985 estimated

Table 3-6. Presidential Requests and Congressional Appropriations for the Defense Budget, Fiscal Years 1980–85

Billions of dollars unless otherwise noted

	Total obligational authority					
Item	*1980*	*1981*	*1982*	*1983*	*1984*	*1985*
Presidential request	135.5	158.7	222.2[a]	258.0	274.1	305.7
Congressional appropriation	141.5	175.3	210.4	238.7	258.2	285.3
Change in amount	+6.0	+16.6	−11.8	−19.3	−15.9	−20.4
Change as percent of request	+4.4	+10.5	−5.3	−7.5	−5.8	−6.7

Sources: *Department of Defense Annual Report, Fiscal Year 1980*, p. 21; *Fiscal Year 1981*, p. 16; *Fiscal Year 1983*, p. I-4; *Fiscal Year 1984*, p. 61; *Fiscal Year 1985*, p. 64; and Department of Defense, *National Defense Budget Estimates for FY 1986*, p. 64.

a. The original Carter request was for $196.4 billion. Reagan proposed adding $25.8 billion to this amount. See table 3-5.

national defense outlays were at their highest real level since fiscal 1969 (the second most costly year of the war in Southeast Asia) and higher than at any other time in postwar, peacetime history.

Other statistics are equally impressive (table 3-7). National defense outlays increased from 5.5 percent of GNP in fiscal 1981 to 6.6 percent in fiscal 1985. Their share of total federal outlays went from 23.2 percent to 26.5 percent during the same four years. And while national defense already constituted 67 percent of all the goods and services purchased by the federal government in fiscal 1981, its share of the total rose to 74 percent by fiscal 1985 (table 3-8). Real authority for defense procurement, perhaps the most striking datum of all, went up by more than 100 percent between fiscal 1980 and fiscal 1985 and 61 percent during the first four Reagan budget years as the administration tried to make up for what had been seen as serious underinvestment during the 1970s. By contrast, real growth in the operation and maintenance account between fiscal 1980 and fiscal 1985 was 37.1 percent, and 13.3 percent in the military personnel account. Table 3-9 shows the growth in these three categories.

The strategic nuclear forces were the most strongly favored by this growth (table 3-10). Between fiscal 1980 and fiscal 1985 procurement for them rose by 182 percent in real terms. The conventional forces received more budget authority but, in the same years, managed real growth in procurement of only 87 percent. Overall the Defense Department, by its emphasis on procurement, was obtaining early congressional approval of an across-the-board modernization of the armed forces, and a rapid modernization at that.

Table 3-7. Outlays as a Percent of Gross National Product and Federal Spending, Fiscal Years 1977–85

Year	*Percent of GNP*		*Percent of federal spending*	
	Defense	*National defense*	*Defense*	*National defense*
1977	5.1	5.2	23.4	23.8
1978	4.9	5.0	22.5	22.8
1979	4.9	4.9	22.8	23.1
1980	5.2	5.2	22.5	22.7
1981	5.4	5.5	23.0	23.2
1982	6.0	6.1	24.5	24.9
1983	6.4	6.5	25.4	26.0
1984	6.2	6.4	25.9	26.7
1985	6.4	6.6	25.7	26.5

Source: Department of Defense, *National Defense Budget Estimates for FY 1986*, pp. 120–21.

Table 3-8. Federal Transactions in the National Income and Product Accounts, Calendar Years 1977–85

Billions of dollars unless otherwise noted

Year	*Purchases of goods and services*[a]		*National defense as percent of total*
	National defense	*Nondefense*	
1977	91.4	48.4	65.4
1978	97.8	52.6	65.0
1979	108.2	55.9	65.9
1980	126.0	63.3	66.6
1981	147.0	71.4	67.3
1982	173.0	77.6	69.0
1983	196.7	76.5	72.0
1984	215.4	69.8	75.5
1985[b]	241.5	85.3	73.9

Source: *Special Analyses, Budget of the United States Government, Fiscal Year 1986*, p. B-27.

a. Other outlays according to this accounting method consist of transfer payments, grants-in-aid, interest payments, and subsidies.

b. Estimated.

Another less noticed area of growth was in the balances of unexpended budget authority from prior-year defense budgets. Those balances stood at $92.1 billion at the end of fiscal 1980. By the end of fiscal 1985 they had grown to $243.4 billion. Real growth during the five years was 98.6 percent, or about the same percentage by which real defense investment increased. However, the unobligated share of the balances rose by only

Table 3-9. Real Growth in the Accounts for Military Personnel, Operation and Maintenance, and Procurement, Fiscal Years 1980–85

Billions of 1986 dollars

	Budget authority		
Year	*Military personnel*[a]	*Operation and maintenance*	*Procurement*
1980	48.0	58.0	50.6
1981	48.9	63.3	63.2
1982	50.6	68.6	79.1
1983	51.8	70.9	93.2
1984	53.2	74.6	95.1
1985	54.4	79.5	101.6
Real growth, 1980–85	13.3	37.1	100.1

Source: Department of Defense, *National Defense Budget Estimates for FY 1986*, pp. 86, 89.
a. Does not include military retired pay.

Table 3-10. Real Growth in Total Obligational Authority for the Strategic and General Purpose Forces, Fiscal Years 1980–85

Billions of 1986 dollars unless otherwise noted

Item	*1980*	*1985*	*Real growth (percent)*	*Average annual real growth (percent)*
Total				
Strategic forces[a]	32.0	50.5	57.8	9.6
General purpose forces[b]	144.8	212.7	46.9	8.0
Procurement				
Strategic forces	6.5	18.3	181.5	23.0
General purpose forces	35.9	67.1	86.9	13.3

Sources: Department of Defense, *National Defense Budget Estimates for FY 1986*, pp. 61, 69; and author's estimates.
a. Consists of program I, and a share of programs VI, VII, VIII, IX, and X.
b. Consists of programs II, IV, V, and a share of programs VI, VII, VIII, IX, and X.

60.0 percent in real terms, while the obligated part more than doubled. Perhaps more important, this growth had a major impact on the composition of national defense outlays (table 3-11). In fiscal 1980, the expenditures resulting from the balances of prior-year authority amounted to 27 percent of total national defense outlays. By fiscal 1985 the prior-year portion had risen to 36 percent and was certain to grow still further. In other words, short of congressional decisions to rescind some of the balances of prior-year authority, a large and increasing percentage of

Table 3-11. Composition of National Defense Outlays, Fiscal Years 1980–85
Billions of dollars unless otherwise noted

Item	*1980*	*1981*	*1982*	*1983*	*1984*	*1985*
Outlays from current-year budget authority	97.5	116.1	128.4	141.6	147.9	162.5
Outlays from prior-year budget authority	36.5	41.4	56.9	68.3	79.5	91.3
Total national defense	134.0	157.5	185.3	209.9	227.4	253.8
Outlays from prior-year budget authority as a percent of total national defense outlays	27.2	26.3	30.7	32.5	35.0	36.0

Source: *Budget of the United States Government, Fiscal Year 1986*, pp. 9-44, 9-48.

both defense and national defense outlays has become uncontrollable. Even if there were no further appropriation of budget authority, the Defense Department would still have more than $240 billion left to spend.

Goals

The Reagan administration, having successfully taken another upward step from the average of the previous twenty-five years during its first four years in office, sought to continue the growth in both the defense and the national defense accounts for the next five years (fiscal 1986 through fiscal 1990). In February 1985, Caspar W. Weinberger, the secretary of defense, proposed a defense budget of $313.7 billion in budget authority for fiscal 1986, a total that would rise to $477.7 billion by fiscal 1990. The cumulative amount over the five-year planning period would add up to $1,985.8 billion. Budget authority requested for national defense was similarly impressive. It would start at $322.2 billion in fiscal 1986 and increase to $488.1 billion by fiscal 1990, for a cumulative total of $2,034.0 billion. Nominal growth (with inflation included) would be over 65 percent for the five-year period and average annual growth would be in the neighborhood of 11 percent. Real growth, given the assumptions used about future inflation, would be more modest: only 37.3 percent for the defense account, or an average of 6.5 percent a year. Outlays would increase a little more rapidly because of the higher rates of growth in budget authority during the previous four years; the 051 account would rise by 38 percent, the 050 account by slightly more than 37

Table 3-12. Outlays as a Percent of Gross National Product and Federal Spending, Fiscal Years 1986–90[a]

	Percent of GNP		*Percent of federal spending*	
Year	*Defense*	*National defense*	*Defense*	*National defense*
1986	6.6	6.8	28.5	29.3
1987	6.9	7.1	30.4	31.3
1988	7.1	7.3	31.8	32.7
1989	7.2	7.4	33.6	34.5
1990	7.4	7.5	35.2	36.0

Sources: *Budget of the United States Government, Fiscal Year 1986*, p. 9-13; and *Special Analyses, Budget of the United States Government, Fiscal Year 1986*, p. A-2.
a. The percentages are based on the assumptions about GNP, federal outlays, defense outlays, and national defense outlays made in the budget submitted on February 4, 1985.

Table 3-13. Real Growth in Investment and Operating and Support Accounts, Fiscal Years 1986–90[a]

Billions of 1986 dollars unless otherwise noted

	Budget authority		
Item	*1986*	*1990*	*Percentage real growth*
Investment[b]	156.4	214.7	37.3
Operating and support[c]	157.3	192.0	22.1
Total	313.7	406.7	29.6

Sources: *Historical Tables, Budget of the United States Government, Fiscal Year 1986*, pp. 5.1(2), 5.1(3); and Department of Defense, *National Defense Budget Estimates for FY 1986*, p. 48.
a. Includes military retired pay.
b. Consists of procurement; research, development, test, and evaluation; military construction; and family housing accounts.
c. Consists of military personnel, operation and maintenance, and other accounts.

percent. It was estimated, as a consequence, that by fiscal 1990 national defense expenditures would amount to 7.5 percent of GNP and a rousing 36.0 percent of federal outlays (table 3-12).

The future real growth in budget authority for defense might not be as impressive as it had been during the four previous fiscal years. Nevertheless, real investment was programmed to rise from $156.4 billion in fiscal 1986 to $214.7 billion in fiscal 1990, a nominal increase of more than 60 percent and a real increase of more than 37 percent (table 3-13). Among the principal beneficiaries of this continued surge, as listed in table 3-14, were to be the Navy's power projection forces ($116.7 billion), the Air Force's air superiority and interdiction capabilities ($58.5 billion), nuclear bombers and cruise missiles designed to penetrate

Table 3-14. Proposed Budget Authority for Major Missions, Fiscal Years 1986–90
Billions of dollars

Mission	*Five-year total*
Power projection (carrier battle groups and amphibious forces)	116.7
Air superiority and interdiction	58.5
Penetration of Soviet air defenses	46.5
Prompt hard-target-kill	45.8
Antisubmarine warfare	33.8
Space and antisatellite warfare	32.7
Ground combat	24.9
Intercontinental mobility	16.1
Field army defense	12.1
Tactical nuclear warfare	8.1
Close air support	6.7
Continental air defense	6.3
Tactical mobility	3.9
Total	412.1

Sources: Department of Defense, "Selected Acquisition Reports as of December 31, 1984," press release, April 9, 1985; and author's estimates.

Soviet air defenses ($46.5 billion), ballistic missiles—both land-based and submarine-launched—all with a high probability of destroying very hard targets ($45.8 billion), and the collection of weapons and research programs related to attacks on orbiting satellites and strategic defense, or Star Wars ($32.7 billion). Thus, although the surge in defense appropriations was scheduled to moderate somewhat, it was clearly intended that the weapons modernization program, begun so aggressively during the early 1980s, would continue at least through the end of the decade.

CHAPTER FOUR

DEBATING POINTS AND FORCE PLANNING

A SOMEWHAT jaundiced Englishman once wrote that one disadvantage of being a hog is that at any moment some blundering fool may try to make a silk purse out of your wife's ear. The president's defense budget is now undergoing that experiment as Congress becomes more reluctant to validate the administration's latest defense requests. Assuming a continuation of the current congressional mood, neither the 051 nor the 050 account will be permitted any real growth in fiscal 1986 and may suffer a real decline if some or all of the allowance for inflation (calculated at around $10 billion) is denied, as the House budget resolution originally proposed. Admittedly, both the House and the Senate intend to be slightly more generous for the next two fiscal years: they project 3 percent real growth in budget authority for national defense in each year. But that is a far cry from the president's original request in his fiscal 1986 message for a 5.9 percent real increase in fiscal 1986, followed by 8.2 percent in 1987, and 8.8 percent in 1988.

Each of the three previous budgetary buildups of any consequence in the postwar period lasted three years, and wars occasioned two of them. The most recent buildup in budget authority began in fiscal 1976 and has now lasted nine years, although six of the nine were, in effect, spent in bringing national defense outlays back to their long-run average of around $213 billion a year. Only since fiscal 1982 have real outlays risen significantly above this average. Table 4-1 shows these periods of real postwar growth.

Even if fiscal 1985 does not turn out to be the highwater mark in budget authority, it seems reasonably evident that the advocates of resistance to further rapid real growth in defense now outnumber the proponents of additional large increases. The persistent and swollen federal deficit and the demand that, as a matter of equity, Defense share

Table 4-1. Periods of Real Growth in Defense Budget Authority and Outlays, Selected Fiscal Years, 1949–85

Billions of 1986 dollars

Year	*Budget authority*	*Outlays*	*Year*	*Budget authority*	*Outlays*
1949	75.5	79.1	1975	177.7	177.5
1950	88.5	77.6	1976	185.8	171.8
1951	248.8	119.3	1977	196.0	174.9
1952	325.3	216.8	1978	193.0	176.0
1953	264.3	234.3	1979	192.7	183.1
1961	185.6	187.1	1980	196.9	188.7
1962	212.6	201.6	1981	221.9	197.7
1963	215.2	205.0	1982	249.0	213.1
1965	196.8	185.0	1983	267.8	228.6
1966	236.4	204.7	1984	279.5	238.3
1967	256.7	241.1	1985	296.1	256.2
1968	260.9	263.1			

Source: Department of Defense, Office of the Assistant Secretary of Defense (Comptroller), *National Defense Budget Estimates for FY 1986*, pp. 84–86, 93–95.

in the burden of reducing it have been the most powerful stimulants to the resistance. Not far behind them in impact has been the drumbeat of stories about waste, fraud, and abuse in the relationship between the Defense Department and its contractors (as implied by the high prices for items listed in table 4-2). There is also a sense in some quarters that the administration has consistently overstated the threat from the Soviet Union, the decade of neglect during the 1970s, the so-called window of vulnerability, and the need for bargaining chips and demonstrations of resolve in negotiations with the USSR. The pace of funding these programs and the multiplicity of new weapons programs have aroused a growing uneasiness. More and more members of Congress, quite apart from the scandals surrounding the procurement process and the overpriced items, have become persuaded that the country is getting a poor return on the large investment it has been making in defense.[1] A small number, mostly senators, have even begun to wonder whether there is any longer a significant relationship among objectives, strategy, forces, and budgets.

All in all this should be enough of a coalition to halt or slow the real growth of the defense budget and even change its allocation. Whether

1. David B. Packard, chairman of the Presidential Commission on Defense Management, has echoed congressional concern. See Bill Keller, "Panel Set to Propose Major Overhaul for Pentagon," *New York Times*, October 11, 1985.

Table 4-2. Examples of High-Priced Items
Dollars

Item	*Price*
Lunch-pack refrigerator	16,571
Coffee brewer	7,622
Landing gear ground lock (for E-2C aircraft)	2,710
Plastic stool leg cap	1,118
Plastic ashtray (for E-2C aircraft)	659
Toilet-seat cover	640
Hammer	435
Socket wrench (to adjust ejection seat on F-14 aircraft)	400

Sources: Editorial, *New York Times*, June 2, 1985; and James Reston, "Farewell to the Jeep," *New York Times*, June 2, 1985.

its members will remain sufficiently cohesive to press the attack, and whether their arguments have sufficient merit to win the day, is another matter.

Defense and the Deficit

Few knowledgeable people any longer question the negative effects of the federal budget deficit on the American and international economies. Even fewer doubt that, subject always to short-run economic and international conditions, it would be desirable to reduce the deficit within a few years from approximately 6 percent of GNP to 2 percent or lower. Exactly where and how that reduction is to be made is, of course, a more contentious subject.

Defense has become an obvious candidate for cuts in outlays, though for reasons that have little to do with the merits of its case. It looks eminently ripe for reduction not only because it is so large, but also because it has grown so rapidly in recent years and currently has so few defenders. Even so, recovering substantial outlays from defense and, at the same time, maintaining a balanced defense capability will not be easy. Under existing circumstances, unless future budget authority for pay and readiness is cut substantially, outlays are likely to fall quite gradually at first, partly because so much of the spending will come from the prior-year balances of unspent authority, and partly because reducing budget authority in accounts such as procurement and military construction will produce very small savings in outlays during the first year of the cuts.

Proposed legislation could have more draconian effects. Passage of the Gramm-Rudman-Hollings amendment to wipe out the federal deficit by 1991 would force Congress and the president either to agree to tax increases or to cut defense outlays by as much as $25 billion in 1987 alone. But because prior-year contracts would not be affected by the amendment, the main burden of the reductions would probably fall on pay and readiness. Thus, the price for restraint imposed in this manner could well be a great deal of new military equipment combined with a shortage of funds to operate and maintain it.

These prospects may seem to suggest that Congress should exempt defense from further cuts, reduce its share of the cuts to much less than the 50 percent proposed in the deficit-cutting amendment, or find other ways to reduce the deficit. Alternatives to cuts in defense outlays obviously exist. Nondefense programs can be reduced still further, as the president has proposed, although opposition to additional major surgery on this portion of the national anatomy appears to be increasingly strong. Despite the president's resistance, taxes can be raised as a substitute for additional reductions in national defense. Some combination of cuts in nondefense programs and increases in taxes might be feasible and eventually may become essential.

It is important, on the other hand, to recognize that many of the arguments both for and against defense cuts have little to recommend them. It is superficial, for example, to claim that defense programs, unlike their nondefense counterparts, are strictly a response to external conditions and therefore cannot be cut in response to economic pressures. Part of what the United States does about defense is undoubtedly conditioned by foreign military capabilities—friendly as well as hostile—and by factors (such as geography and weather) over which the United States may have no control. But the president and the secretary of defense do control a number of the other variables that shape the defense budget. They are largely responsible for setting the objectives the armed forces are designed to achieve. Explicitly or implicitly they must decide on the level of confidence they want in being able to reach these objectives. They must ascertain the opportunity costs of investments in defense while considering other national objectives such as ensuring a healthy, stable, and growing economy. Defense no doubt should be treated on its merits. But there are no absolutes here, no certainty of success or failure, only probabilities and risks. Lower probabilities and higher risks than originally sought may become acceptable if the federal deficit is seen as the greater threat.

Waste, Fraud, and Abuse

Scandals in defense procurement, by contrast, are more an excuse than a serious basis for major reductions in the defense budget. Some outrages are the result of an attitude toward expense accounts that is pervasive well beyond the defense industry. Others undoubtedly come from creative accounting and a propensity to charge overhead costs to items that ordinarily attract little attention. The case for putting an end to these practices wherever possible is impeccable even though the costs of doing so could easily equal or exceed the savings. Conceivably, with defense procurement already about $100 billion a year, more rigorous auditing could save as much as 3 percent of the total, or a nontrivial $3 billion. But it is reasonably evident that even double or triple these savings would not contribute materially to a reduction of the deficit. Nor will reforms in the defense procurement process prevent unwise decisions from being made about what is to be procured.

The Threat

The charge that the administration, whether out of ignorance or design, has exaggerated the need for its large increases in the defense budget, the centerpiece of which has been the growth in procurement, is not difficult to sustain. Much has been made, for example, of what has been described as the unprecedented Soviet military buildup, shown in dollars and compared with U.S. outlays in figure 4-1. Very little has been made of the relatively low base from which the current buildup began, or of how much of it went to restoring the cuts made by Khrushchev in the late 1950s and early 1960s, constructing a modern strategic nuclear capability, and nearly tripling the forces deployed in the Far East after the falling out with China. The pace of the Soviet buildup—advertised to have been on the average a 4 percent real increase each year—dropped off to 3 percent a year in 1974 as growth in the Soviet economy began to stagnate. The increases then stayed at 2 percent from 1978 through 1981, and quite probably at least a year after that (figure 4-2). Moreover, during those six or seven years, all the increases took place in the operating and support accounts; procurement did not grow at all.

Despite the slowdown, despite the preoccupation with China and with its southern border, the Soviet Union has certainly developed the

Figure 4-1. U.S. Defense Outlays and Estimated Dollar Cost of Soviet Defense Activities, 1951–81

Source: *Department of Defense Annual Report to the Congress, Fiscal Year 1983*, p. I-20.

capability to threaten U.S. interests. And it continues to create uncertainty and suspicion about its intentions not least because of the secrecy of its decisions and activities. It is also true that in the wake of the war in Southeast Asia—while the Soviet Union continued its military buildup and modernization—the United States had reduced its real outlays for defense to $171.8 billion (fiscal 1986 dollars) by fiscal 1976, the lowest they had been since the outset of the Korean War (table 4-3). U.S. real defense investment between fiscal 1971 and 1975 also fell well below the levels maintained in the 1950s and 1960s (table 4-4).

The Decade of Neglect

Whether the declines in the 1970s warrant the title of a decade of neglect is a more arguable matter. However the outcome of the war in

Figure 4-2. Growth in Soviet Defense Program, 1970–81
Billions of 1981 dollars (estimated)

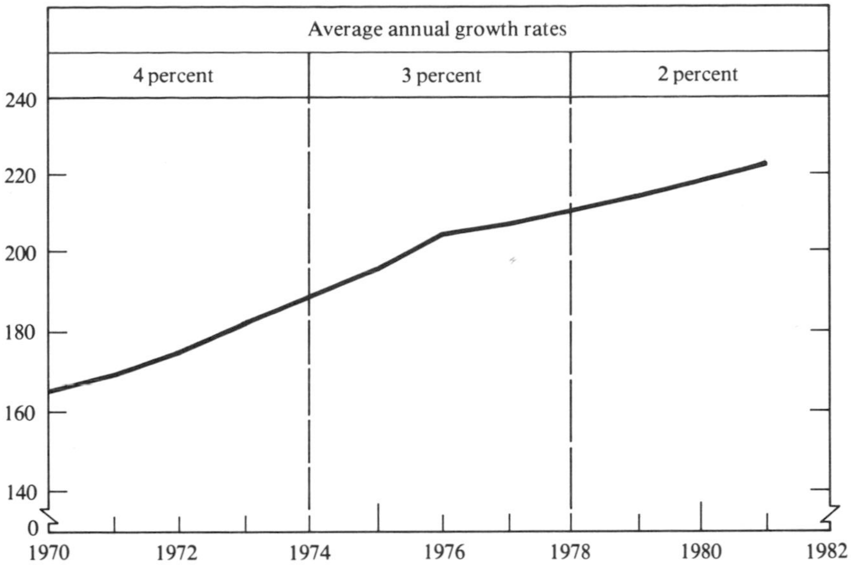

Average annual growth rates
4 percent
3 percent
2 percent
240
220
200
180
160
140
0
1970
1972
1974
1976
1978
1980
1982

Source: *Allocation of Resources in the Soviet Union and China—1983,* Hearings before the Subcommittee on International Trade, Finance, and Security Economics of the Joint Economic Committee, 98 Cong. 1 sess. (Government Printing Office, 1984), pt. 9, p. 101.

Southeast Asia should be characterized, U.S. armed forces did not suffer any major defeats in the field, and they withdrew from the theater in a deliberate and orderly fashion. As had been true in previous wars the conflict was accompanied by the rapid production of modern weapons, equipment, and supplies, particularly for the Army and Air Force, much of which the two services retained. As a consequence, a major investment program was not needed immediately. During the latter half of the 1970s, funds for research and development, military construction, and procurement became as large in real terms as they had been during the second term of the Eisenhower administration. Indeed, between fiscal 1974 and 1977 the Ford administration funded essentially as many Trident ballistic missile submarines, tanks, combat aircraft, and warships as the Reagan administration did during its first four years. And while the priorities of the Carter administration were somewhat different, as were its allocations, the general pattern was the same (table 4-5).

None of this is to say that the armed forces escaped untouched from the war. They did not. As they withdrew from Vietnam and the all-volunteer force replaced conscription, active-duty personnel fell from a peak of 4.6 million to fewer than 3.0 million, while total Defense

Table 4-3. The Defense Budget in the Decade of Neglect, Fiscal Years 1971–81
Billions of dollars

President and year	*Current dollars*		*1986 dollars*	
	Budget authority	*Outlays*	*Budget authority*	*Outlays*
Nixon				
1971	71.2	74.5	203.8	213.7
1972	75.0	75.1	197.5	198.9
1973	77.6	73.2	188.5	181.8
1974	81.0	77.6	183.2	177.9
Ford				
1975	85.7	84.9	177.7	177.5
1976	95.5	87.9	185.8	171.8
1977	108.3	95.6	196.0	174.9
Carter				
1978	115.3	103.0	193.0	176.0
1979	125.0	115.0	192.7	183.1
1980	142.6	132.8	196.9	188.7
1981	178.4	156.1	221.9	197.7

Source: Department of Defense, *National Defense Budget Estimates for FY 1986*, pp. 85–86, 94–95.

Table 4-4. Defense Investments by Five-Year Periods, Fiscal Years 1951–85
Billions of 1986 dollars

Period	*Budget authority*[a]
1951–55	498.0
1956–60	332.0
1961–65	417.5
1966–70	451.4
1971–75	296.0
1976–80	344.3
1981–85	588.7
Total	2,927.9

Source: Department of Defense, *National Defense Budget Estimates for FY 1986*, pp. 87–89.
a. Consists of procurement; research, development, test, and evaluation; and military construction accounts.

Department manpower (military and civilian) dropped from nearly 5 million to just above 3 million, not including reserve components (table 4-6). The Army lost six divisions and the Air Force three tactical fighter wings from their prewar force structure, but both services were returning to that baseline by the end of the decade. Perhaps the most dramatic decline occurred in the Navy, which, in the net, retired nearly 300 ships from its active fleet despite a shipbuilding program that, on the average,

Table 4-5. Acquisition of Major Weapons Systems, Fiscal Years 1974–85
Number of items

Weapons system	*1974–77*	*1978–81*	*1982–85*
Strategic and tactical nuclear missiles	826	1,192	2,284
Trident submarines	5	4	3
Tanks	3,241	2,524	3,235
All other vehicles	3,691	4,420	7,107
Surface-launched tactical missiles	173,295	78,199	79,860
Air-launched tactical missiles	29,101	19,164	42,047
Combat aircraft	1,425	1,841	1,482
Airlift aircraft	111	136	165
Trainer aircraft	227	117	114
Helicopters	561	612	1,055
Major warships	29	16	29
Other warships	21	28	22
Auxiliary ships	11	15	29

Source: Author's estimates based on *Department of Defense Annual Report*, Fiscal Years 1974–85.

Table 4-6. Defense Personnel and Force Structure, Fiscal Years 1964, 1968, 1973

Item	*1964*	*1968*	*1973*
Defense personnel (in thousands)			
Active-duty military	2,685	3,547	2,252
Reserve components (in paid status)	1,048	1,001	972
Direct-hire civilians	1,035	1,287	998
Strategic forces			
ICBMs	708	1,018	1,054
Fleet ballistic launchers	336	656	656
Bomber squadrons	78	34	30
Fighter interceptor squadrons (active)	40	26	7
Air defense batteries	107	83	21
General purpose forces (active)			
Army divisions	16⅓	19	13
Marine Corps divisions	3	4	3
Air Force fighter-attack wings	21	25	22
Navy fighter-attack wings	15	15	14
Marine Corps fighter-attack wings	3	3	3
Attack carriers	15	15	14
Antisubmarine warfare carriers	9	8	2
Nuclear attack submarines	19	n.a.	60
Amphibious warfare ships	133	n.a.	66
Intercontinental airlift aircraft (C-141, C-5A)	6	266	351
Troopships, cargo ships, tankers	101	130	53

Sources: *Department of Defense Annual Report, Fiscal Year 1975*, pp. 236–37; *Fiscal Year 1983*, p. A-3; and author's estimates.
n.a. Not available.

was funding 15 new ships a year. However, much of this reduction was inevitable. The ships being retired were of World War II vintage and had reached the end of their useful service lives. No one (at least outside the Navy) believed that they should be replaced on a one-for-one basis, considering that the new additions to the fleet were larger, more capable, and a great deal more expensive in real terms. It would, of course, have been possible to build a greater number of smaller and cheaper ships to keep up the size of the fleet, as some senior naval officers proposed. However, the Navy ended up choosing a smaller number of large and costly ships, although it also tried to stabilize the size of the fleet at 600 ships, a number the chief merit of which was that it was round.

The Window of Vulnerability

Belief in the relentless Soviet buildup and the U.S. decade of neglect contributed strongly to the conviction of the president, quickly adopted by his secretary of defense, that the United States had fallen behind seriously in the military competition with the USSR. In addition, Reagan seemed genuinely to believe that the slippage had made the United States immediately vulnerable to attack, though Weinberger suggested that the mid-1980s would be the period of maximum peril.[2] The main reason for this window of vulnerability apparently was the estimate that U.S. land-based intercontinental ballistic missiles (ICBMs) were declining in survivability. But the president also insisted that the USSR had taken the lead over the United States in every relevant measure of military power.[3]

That these fears lacked serious foundation was demonstrated not only by the Central Intelligence Agency (CIA) data about the less than "relentless" Soviet buildup and the facts about the U.S. "decade of neglect," but also by independent analysis and the administration's own programs. In 1983 the Commission on Strategic Forces, known as the Scowcroft Commission, established by the president, concluded that the U.S. ICBM force was indeed becoming vulnerable because of the large throw-weight and increasing accuracy of Soviet ICBMs. But the commission also agreed that the overall ability of the U.S. strategic

2. *Department of Defense Annual Report to the Congress, Fiscal Year 1983,* p. I-39.

3. "Arms Reduction and Deterrence: Address to the Nation," *Weekly Compilation of Presidential Documents,* vol. 18 (November 18, 1982), pp. 1516–21.

nuclear forces to survive a surprise attack and retaliate against a wide range of targets (military as well as economic) remained impressive enough to constitute a highly effective deterrent to nuclear attacks. The administration, for its part, showed a certain insouciance about the window of vulnerability. It took none of the emergency measures it could have adopted to increase its survivable, second-strike nuclear capability, such as placing more heavy bombers on day-to-day ground alert and more ballistic missiles in submarines on station and within reach of Soviet targets, even though these higher alerts could have increased by as many as 1,700 the number of deliverable warheads.

Still more significant, the large investment programs launched by the Defense Department, presumably with the approval of the secretary of defense and the president, will not appreciably narrow the gap between the USSR and the United States according to the measures used by Reagan to describe and dramatize the state of the military competition. Despite these programs and their costs the Soviet Union will continue to have more ICBMs and ballistic missile submarines than the United States (short of a major arms control agreement equalizing the numbers during the next seven years), more men under arms, more tanks, armored fighting vehicles, combat aircraft, and warships. In the circumstances it must be assumed either that there is something badly wrong with the programs of the Defense Department or that, if the measures used by the president are misleading about the future, they must have been equally misleading about the past.

Return on Investment

Whatever may be the case—and it could be a combination of the two—it is clear why the return on the administration's investment in defense has become an issue in Congress and elsewhere. The issue is not easy to resolve, mostly because there is no agreement on what should be defined as return on investment or how it should be measured. The Congressional Budget Office has attempted to circumvent these difficulties by two ingenious methods. First, in determining what constitutes the return on investment the office looks at what the three military services regard as the four main determinants of defense capability: force structure, modernization, readiness, and sustainability. Second, as a measure of the return on investment in these four areas, the office

Table 4-7. Defense Personnel and Force Structure, Fiscal Years 1976, 1980, 1985

Item	*1976*	*1980*	*1985*
Personnel (in thousands)			
Active military	2,081	2,040	2,152
Reserve components (selected reserve)	823	861	1,077
Direct-hire civilians	960	916	1,002
Strategic forces			
ICBMs	1,054	1,052	1,023
Fleet ballistic launchers	656	576	640
Bombers[a]	452	376	298
Fighter interceptors (active)[a]	141	127	90
General purpose forces (active)			
Army divisions	16	16	17
Marine Corps divisions	3	3	3
Air Force fighter-attack aircraft[a]	1,608	1,680	1,758
Navy fighter-attack aircraft[a]	976	894	956
Marine Corps fighter-attack aircraft[a]	422	422	417
Attack carriers	13	13	13
Nuclear attack submarines	63	74	96
Amphibious warfare ships	62	66	61
Cruisers, destroyers, frigates	177	178	209
Intercontinental airlift aircraft (C-141, C-5A)[a]	304	304	304
Cargo ships, tankers	31	35	40

Sources: *Department of Defense Annual Report to the Congress, Fiscal Year 1978,* p. C-5; and *Department of Defense Annual Report, Fiscal Year 1986,* pp. 155, 300, 303–305.

a. These are primary aircraft authorized (PAA); they do not include aircraft acquired for attrition, maintenance, or training.

compares the accomplishments of the Carter administration with those of the Reagan administration during its first four years in office.

The application of this methodology provides several insights. For example, Reagan has added a little more to the U.S. force structure (consisting of such units as missiles, bombers, divisions, wings, ships) than Carter did, but the comparison is among a variety of disparate units and the costs of the changes were not determined. Carter held the number of active-duty and ready-reserve Army divisions at 24; Reagan added 1. Carter added 8 Air Force fighter-attack squadrons; Reagan added another 2. Under Carter the number of Navy fighter-attack squadrons fell from 75 to 70; under Reagan it went back up to 72. Carter also presided over a reduction in the Navy's deployable battle force ships from 484 to 479; Reagan was able to raise them from 479 to 523. However, all but 5 of the 44 additional ships were funded by the Ford and Carter administrations. Major changes in the forces between fiscal 1976 and 1985 are shown in table 4-7.

Table 4-8. Modernization of Military Equipment under Presidents Carter and Reagan, Fiscal Years 1977–85

Number of units; costs in billions of 1985 dollars

Equipment	*1977–80*[a]	*1982–85*[a]	*Percent change*
Aircraft, fixed wing			
Combat	1,745	1,482	-15.1
Airlift	144	165	14.6
Trainer	113	114	0.9
Aircraft, rotary	587	1,055	79.7
Total aircraft	2,589	2,816	8.8
Total cost	43.3	75.9	75.4
Missiles			
Strategic and theater nuclear	627	2,284	264.3
Tactical air-launched	19,999	42,047	110.2
Tactical surface-launched	96,082	79,860	-16.9
Total missiles	116,708	124,191	6.4
Total cost	15.0	28.7	91.2
Ships			
Trident submarines	4	3	-25.0
Major warships	15	29	93.3
Other warships	29	22	-24.1
Auxiliaries	13	29	123.1
Total ships	61	83	36.1
Total cost	28.9	44.2	53.0
Tanks and combat vehicles			
Tanks	2,762	3,235	17.1
All other vehicles	5,194	7,107	36.8
Total vehicles	7,956	10,342	30.0
Total cost	6.2	15.3	147.4

Source: Congressional Budget Office, "Defense Spending: What Has Been Accomplished," Staff Working Paper, April 1985, p. 13.

a. Units were ordered, but not necessarily delivered, during these periods. They do not include classified programs.

The findings of the Congressional Budget Office about modernization are more detailed (table 4-8). Reagan, in his first term, funded 6.4 percent more missiles (strategic, theater nuclear, and conventional) than Carter, but paid 91.2 percent more in constant dollars. Reagan bought 8.8 percent more aircraft (fixed wing and rotary), but they cost him 75.4 percent more. Reagan managed to acquire 30.0 percent more tanks and other combat vehicles, but he had to pay 147.4 percent more for them. He did best with ships: 36.1 percent more of them were funded while the cost was only 53.0 percent higher.

The Congressional Budget Office was not able to adopt the same approach in looking at readiness. But by using the standard service measures, it found that the armed forces are now substantially more prepared for combat than they were in 1980. The biggest improvements have come in the quality of the enlisted personnel and the rates of their reenlistment. How much of this has been because of pay raises, and how much because of the major recession of 1982–83, the office does not say. Recent data about recruitment suggest, however, that the quality and even the number of potential enlistees tend to be sensitive to the condition of the economy.

Improvements have also taken place in what are known as the mission-capable rates of equipment—that is the percentage of the weapons platforms on hand deemed capable of performing their wartime assignments. All of the main systems, except for the Navy's fighter-attack aircraft, now have mission-capable rates of 70 percent or better. By this measure, in fact, most of them have achieved levels of readiness that now approximate or exceed the goals set by the services (table 4-9). Training, however, does not appear to have increased greatly. Funding for it has risen only modestly, the number of man-years devoted to individual training has remained relatively constant, and the days per year spent on training ground force units, the hours flown per aircrew per month, and the days per quarter steamed by individual ships have increased by small amounts if at all (table 4-10).

The administration has allocated large funds (more than $40 billion in budget authority) to munitions and other matériel and supplies for the war reserve stocks of the armed forces, and the services agree that progress has been made toward their objectives. But only the Army claims to have more than 50 percent of the war reserves it needs. And despite the efforts of the administration, the services estimate that, on the average, they will need another six years at current rates of funding, and more than $70 billion, to reach their expanding goals (table 4-11).

The impressions to be drawn from these comparisons and snapshots are bound to be mixed. The Reagan administration has clearly made progress on modernization, readiness, and force sustainability as the Congressional Budget Office, and for the most part the services, measure it. At first glance, however, the comparisons with the Carter administration suggest that the country is now obtaining a much lower return on its defense investment than it did prior to 1981.

Whether such a judgment would be fair to the Reagan administration is, however, not that certain. The principal if not the only reason why the cost-benefit ratios look so unfavorable to Reagan is because defense

Table 4-9. Percentage of Equipment Capable of Performing Mission, Fiscal Years 1980, 1984

Service and equipment	*Goal*	*1980*	*1984*
Army			
Aircraft	75	66	71
Artillery	90	88	89
Missiles	90	91	94
Tanks	90	86	87
Navy			
Fighter-attack aircraft[a]	68	53	63
Total aircraft[a]	73	59	70
Marine Corps			
Artillery	85	88	89
Missiles	85	94	86
Tanks	85	86	87
Air Force			
Fighter-attack aircraft	74	62	73
Total aircraft	75	66	71

Source: CBO, "Defense Spending," p. 20.
a. Includes Marine Corps aircraft.

Table 4-10. Military Training, Fiscal Years 1980, 1982, 1984

Item	*1980*	*1982*	*1984*
Individual training loads (thousands of man-years)			
Army	78	76	70
Navy	58	64	64
Marine Corps	19	19	21
Air Force	42	44	41
Reserve components	28	38	32
Total Defense Department	236	241	228
Funding for training (billions of 1985 dollars)	11.1	12.6	13.8
Collective unit training			
Annual training days per battalion			
Army	n.a.	161.7	161.9
Marine Corps	n.a.	95.2	100.5
Flying hours per crew per month			
Army	18.8	17.2	16.4
Navy and Marine Corps	24.2	23.7	23.7
Air Force	20.2	21.4	21.5
Air Force tactical aircraft	15.6	n.a.	19.3
Steaming days per quarter			
Deployed fleets	57	58	60
Nondeployed fleets	29	29	28

Source: CBO, "Defense Spending," p. 29.
n.a. Not available.

Table 4-11. War Reserve Stocks of Conventional Munitions[a]

Service	Percent of objective		Funding, 1981–85 (billions of dollars)	Additional cost to reach objective (billions of dollars)	Additional years to reach objective[b]
	1980	1984			
Army	65	77	19.1	15.4	3
Navy[c]	12	22	12.1	25.4	7
Marine Corps	32	44	2.8	3.8	5
Air Force	21	30	11.8	25.8	9
Total	...	...	45.8	70.4	6

Source: CBO, "Defense Spending," p. 22.
a. Consists of ammunition, bombs, and most tactical missiles.
b. At fiscal 1985 spending rates.
c. Includes Marine Corps air munitions.

is now buying more of the more sophisticated weapons systems than it could afford during Carter's administration. In effect, the services are betting heavily on quality rather than numbers, although they would obviously like to have both. The analysis of the Congressional Budget Office does not pretend to and cannot show whether the Defense Department is making a sensible trade between the two. Nor can the analysis take account of changes in the threat, which might justify both more expensive weapons and more of them. Perhaps equally important, the results of the comparisons cannot lead to recommendations about what should be done in light of the seemingly low return on investment. This lack of any operational guidance, other than perhaps to grin and bear it regardless of the costs, is almost inevitable when the analysis is based on inputs rather than outputs, and when the method offers no way of defining military objectives or telling how effectiveness in reaching them changes with cost.

Comparisons of GNP

The approach used by the Congressional Budget Office may not deal with these issues, give any sense of the probability of reaching an objective, or note the risks of failure. But what are the alternatives? One seemingly simple substitute is to determine the percentage of GNP being committed by the Soviet Union to its military effort and use it as the basis for the U.S. defense program. Thus, if the USSR is spending 14 percent of its GNP for military purposes, and its GNP is half that of the

United States, then the United States should plan to commit 7 percent of its GNP to defense. The rather large presumption is that if the resulting commitment matches that of the USSR, it will be at least as wisely allocated, and the capabilities produced by it will be at least as effective. Apart from its simplicity, however, this approach has little to recommend it. Suppose, for example, that Mikhail S. Gorbachev adopts the same formula but, owing to various reforms, is able to induce much more rapid economic growth in the Soviet Union, while the United States remains trapped in a growth recession. Under these conditions the Defense Department will be committed to a 7 percent share of a sluggish GNP, but the Soviet ministry of defense, benefiting from 14 percent of its country's rapidly growing GNP, will have a much larger budget with which to threaten the United States. If U.S. GNP declined in real terms, as it has been known to do, the disparity could become even worse. Such considerations aside, and discarding the possibility that Soviet GNP may be difficult to estimate accurately, the dedication of a certain percent of GNP to defense creates other problems. Regardless of whether the product goes up or down, or whether it changes rapidly or slowly, fixing the percentage does not dispose of the issue of its efficient allocation. Nor does the policymaker gain any sense of what he or she would lose by forgoing some defense resources and capability in favor of nondefense programs.

Budget Comparisons

A second possibility concentrates on the Soviet military budget and its allocation, particularly to investment, as a basis for determining what the United States should do. The assumption is that if, at a minimum, American policymakers mimic the Soviet total as well as its allocation, they will forestall any Russian advantage, presumably on Annie Oakley's axiom that anything you can do I can do better. In principle, this approach has two attractions: it requires little thought, and it seems highly conservative in that it ensures an identical response to whatever the Russians might be doing. In practice, such an approach raises more issues than it resolves. To begin with, the U.S. intelligence community does not really know the costs of the Soviet military establishment. The USSR publishes patently false figures about the magnitude of its defense program and says nothing about its allocation. Moreover, since the Soviet Union lacks anything resembling a market economy, the exchange value of rubles is an inaccurate measure of opportunity costs within Soviet society. When American officials talk about Soviet defense spending, they are referring for the most part to the CIA estimates of what it would cost the United States to own and operate the Soviet military establishment in the U.S. economy at U.S. prices. The CIA makes no bones about what it is doing or the limitations of the methodology. In an effort to demonstrate some of its distortions the agency also estimates, though with much less confidence, what it would cost the USSR to own and operate the U.S. defense establishment in the Soviet economy and at Soviet prices. However, neither technique can fully account for certain fundamental differences between the United States and the USSR. Clearly they have differing conceptions of their interests, what threatens them, and their objectives. The Soviet Union is both a poorer and a more populous country and, in its armed forces as elsewhere, is more likely to use labor as a substitute for scarce capital. The two countries have different geographies, climates, and allies with which to contend. Because of pervasive Soviet secrecy it is extremely difficult to compare Soviet weapons and technology with comparable U.S. capabilities, although samples of both occasionally fall into U.S. hands. However, from the evidence available (which includes the intensive Soviet effort to buy or steal Western technology) it seems reasonably clear that Soviet weapons systems do not for the most part come up to U.S. standards. Because of these and other differences, the United States would never dream of buying the equivalent of the Soviet military establishment. And quite implausibly, even if the United States faced the same threats and had the same interests and objectives, it would probably design something different.

None of this should be taken as advice to dismiss the CIA estimates as irrelevant. Although they undergo periodic revisions as new or old data leak out of the Soviet Union, the estimates are interesting indicators of overall trends. They also raise intriguing issues such as why a U.S. division and its support forces cost three times more than the Soviet equivalent (in U.S. prices), yet is considered only 20 percent more powerful by the U.S. Army.

Comparative Investments

Caspar W. Weinberger, the secretary of defense, has also argued that whatever the merits of comparing the overall defense expenditures of

Figure 4-3. U.S. and Soviet Defense Investments[a]

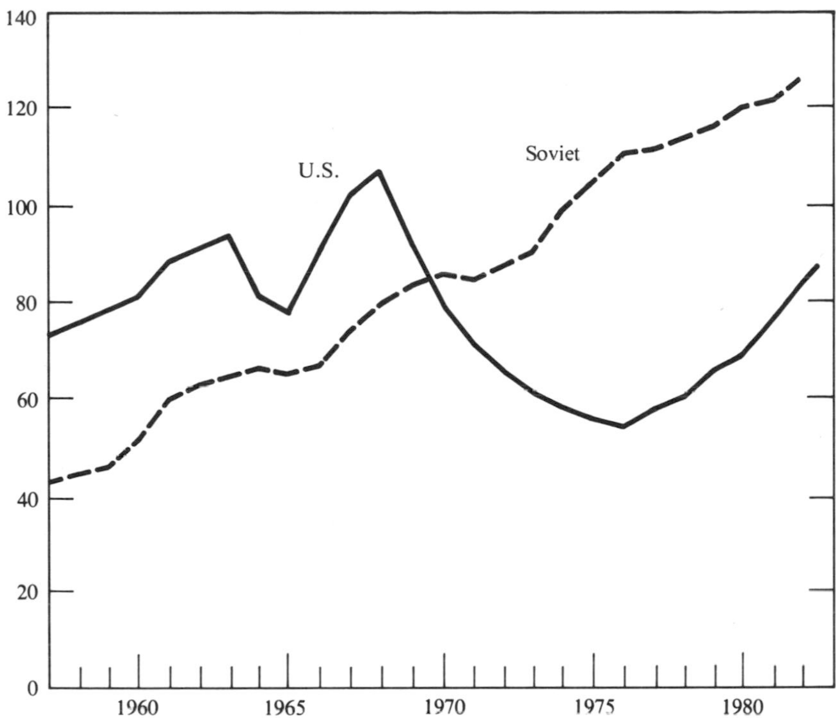

Source: *Department of Defense Annual Report to the Congress, Fiscal Year 1985*, p. 20.

a. U.S. outlays for investment are defined as the sum of procurement; research, development, test, and evaluation; and military construction. Soviet outlays for investment are defined as the estimated dollar cost of reproducing these investments in the U.S. economy.

the two countries, it is noteworthy and alarming that the USSR—according to the CIA estimates—invests nearly twice as much as the United States in research and development, military construction, and procurement (figure 4-3). Indeed, that gap seems to have been a major consideration in the administration's decision to put so much weight on investment in its own budgets. To what extent the United States has now caught up in the investment race has not been revealed. But there is at least a lurking suspicion that the United States is racing with itself. As the CIA is the first to admit, it has great difficulty estimating the precise nature and costs of Soviet research and development. And it does not even attempt to assess its efficiency. The output of the Soviet defense industry is much easier to calculate. But there is bound to be great uncertainty when attempting to cost that output in U.S. prices,

Table 4-12. Potential Soviet Deployment of General Purpose Forces by Theater of Military Operations

Item	Theater of military operations (TVD)							
	Arctic[a]	Atlantic[b]	Northwestern	Western[c]	Strategic reserves	Southwestern[d]	Southern[e]	Far East[f]
Divisions	...	...	10	62	18	26	30	53
Tanks	...	...	1,400	19,680	4,590	6,890	5,200	14,900
Artillery, mortar, multiple rocket launchers	...	...	2,375	15,750	4,170	5,670	6,600	15,200
Tactical aircraft	...	...	225	2,290	150	890	890	1,690
Aircraft carriers	...	...	...	...	...	1	...	2
Principal surface combatants	80	43	...	...	...	74	5	85
Other combatant ships and craft	132	347	...	...	...	235	65	354
Auxiliaries	200	170	...	...	...	150	25	235
Submarines[g]	142	33	...	...	...	33	...	110
Naval aviation	440	270	...	...	...	435	...	500
Naval infantry brigades	1	1	...	...	...	1	...	3

Source: Department of Defense, *Soviet Military Power* (GPO, 1985), pp. 13–15.
a. Consists of the Northern Fleet.
b. Consists of the Baltic Fleet.
c. Does not include 31 divisions and 795 tactical aircraft of East Germany, Poland, and Czechoslovakia.
d. Includes the Black Sea Fleet. Does not include 24 divisions and 345 aircraft of Hungary, Romania, and Bulgaria.
e. Includes the Caspian Sea Flotilla and the Soviet forces operating in Afghanistan.
f. Includes the Pacific Fleet.
g. Does not include 62 ballistic missile submarines.

especially when the United States obtains only limited hands-on access to that output and usually acquires whatever Soviet equipment comes its way some years after it has gone into service. The United States, rather than acknowledging this uncertainty, has the propensity to impute U.S. technology, U.S. sophistication, U.S. effectiveness, and U.S. prices to whatever is under consideration, especially if the Defense Intelligence Agency rather than the CIA is the source of information about a particular Soviet weapon or research and development program. Consequently, the next generation of U.S. weapons must outperform the current generation of U.S. weapons—and by a wide margin—even though when occasionally the Soviet systems are actually acquired, as in the famous case of the MiG-25 (Foxbat) interceptor that escaped to Japan, they rather consistently turn out to be substantially inferior to what the United States has in service.

Additional complications in comparing the defense investments of the two countries arise from the larger size of the Soviet military establishment and the modernization strategy Soviet leaders appear to have adopted. Most estimates put Soviet military manpower at about 4.2 million (not counting border guards and KGB formations), or about twice the number the United States has in uniform. If it is assumed that on the average the USSR capitalizes each of its men as heavily as the United States, and that efficiencies in defense production are about the same—assumptions probably favorable to the USSR—it obviously follows that Soviet defense investment would be about twice as large as that of the United States. Yet it is improbable that all the Soviet forces would manage somewhere to encounter all the U.S. forces or that either would fight the other in key theaters without the company of allies. Indeed, when account is taken of where Soviet forces are deployed (as shown in table 4-12), the likelihood that many of them would or could be tied down along the eastern and southern borders of the USSR, and of the comparatively large investments made by U.S. allies (much larger than the efforts of the Soviet satellites), the investment balance looks a good deal less alarming than is suggested by the usual gross comparisons.

If one uses the CIA methodology, a question arises about whether the Soviet Union cumulatively has invested more in defense than the United States has. The answer depends critically on the year from which the comparison is started. If it is 1970, the Soviet Union is well ahead; if it is 1950, the two sides are about even. The response to this unsettling conclusion has been that whereas the United States made its heaviest investments during the 1950s and 1960s, the USSR undertook a big expansion and modernization program during the latter half of the 1960s and all throughout the 1970s, when the United States was wasting investments on Vietnam and subsequently reducing them during the decade of neglect. Therefore, although cumulative investments may be about equal in real terms, the USSR has by far the larger stock of modern capital goods in its military establishment.

One of the difficulties with this argument is that it ignores the differing ways in which the two countries modernize their respective stocks of weapons and equipment. Like the hare, the United States has historically made investments in fairly short spurts; like the tortoise, the USSR has tended to plod along, gradually replacing its stock of systems. What the Russians produce in this process always appears large and alarming because Soviet forces and their capital stock are indeed large (table 4-13). But once it is appreciated that even if 3,000 tanks a year were produced (and none of them exported), it would take at least fifteen years to replace the current inventory of Soviet tanks, the numbers become less frightening.

Further consolation, if needed, can be drawn from the U.S. practice of upgrading weapons and equipment during their normal service lives, averaging about twenty years. The classic result of this practice is the current B-52 bomber. True, the last B-52 was produced in 1962, so that the fleet is now chronologically old. But the aircraft currently in the fleet have had their wings rebuilt to sustain low-altitude flight, their avionic systems (for navigation, electronic countermeasures, and bombing) have been upgraded several times, and they have been reskinned and reengined as well. The cost of these modifications, in constant dollars, has probably exceeded the original price of the aircraft by a factor of two. The Army has done much the same thing with its tanks, and the Navy is now rebuilding its older aircraft carriers at a cost of $800 million per ship. Many of these costs do not even appear in the investment accounts, since funds for the installation of new equipment usually come from the operation and maintenance accounts. Yet they are just as surely improvements in the capital stock as the procurement of new weapons and equipment.

Whether the United States keeps as much of its matériel combat-ready as the USSR is uncertain: the two countries follow different practices in alerting and operating their forces in peacetime, with American forces much the more active of the two. But it should be evident that investments in capital goods will not be very productive if personnel are not trained to operate and maintain them. A comparison

Table 4-13. Selected Soviet Military Equipment Inventories and Production[a]

Equipment	*Estimated inventory*	*Annual production (five-year average, 1977–81)*	*Years required to replace inventory*
Tanks	52,660	2,020	26
Infantry carriers	60,000	4,280[b]	14
Field artillery	29,000	1,460	20
Multiple rocket launchers	4,000	280	14
Strategic, naval, tactical aircraft	9,706	810	12
Major combatant ships	290	9	32
Minor combatant ships	1,133	28	40
Auxiliaries	780	6	130
Submarines	380	11	35

Sources: Department of Defense, *Soviet Military Power*, pp. 8–9, 13–15, 67, 68; and *Allocation of Resources in the Soviet Union and China—1983*, Hearings before the Subcommittee on International Trade, Finance, and Security Economics of the Joint Economic Committee, 98 Cong. 1 sess. (Government Printing Office, 1984), pt. 9, p. 379.

a. Weapons for export are not included in these totals.

b. Includes between 600 and 800 vehicles imported yearly from Eastern Europe.

of investment accounts, whether on a year-to-year or on a cumulative basis, will capture only a portion of these operating costs, many of which are billed to operation and maintenance.

These defects aside, a comparison of investments provides little guidance to policymakers. At best, if interpreted with some sophistication, it can signal an impending problem; at worst, such a comparison can mislead about the seriousness of the investment "gap." It is of little or no use in telling the policymaker by how much he or she should change the amount of an investment or how to allocate it.

Force Comparisons

What the military calls bean counts—the enumeration and comparison of such factors as personnel, weapons, and equipment, as illustrated in table 4-14—are usually the source of President Reagan's conclusion that the USSR leads the United States in every "relevant measure" of military power. But are bean counts any more relevant than comparisons of overall budgets, investments, and other crude indicators for assessing particular military balances and engaging in efficient force planning?

Certainly the bean counts are an improvement over spending-type measures in that, at the least, they refer to items such as missiles, aircraft, tanks, and ships. It is also true that numbers can defeat quality in combat; Napoleon himself seemed to believe on occasion that God was on the side of the big battalions. But as Horatius would probably have testified

Table 4-14. Defense Intelligence Agency Count of NATO and Warsaw Pact Forces, 1983

Category	*NATO*[a]		*Warsaw Pact*	
	Rapidly deployable	*Fully reinforced*	*Rapidly deployable*	*Fully reinforced*
Strategic forces				
U.S. and Soviet intercontinental ballistic missiles	1,026[b]	...	1,398	...
U.S. and Soviet submarine-launched ballistic missiles	640[b]	...	982	...
U.S. and Soviet bombers	296[b]	...	303	...
Land forces				
Main battle tanks	13,470	17,730	26,900	46,230
Artillery and mortars	11,000	14,700	19,900	38,800
Attack helicopters	560	900	1,135	1,175
Transport and support helicopters	1,960	6,000	1,180	1,375
Armored personnel carriers and fighting vehicles	33,000	39,580	53,000	94,800
Antitank guided weapon launchers	12,340	19,170	18,400	35,400
Division equivalents	88	115	115	192
Combat aircraft				
Fighter-bomber, ground-attack	1,960	...	2,250	...
Interceptor	795	...	4,195	...
Reconnaissance	235	...	585	...
Bombers	...	...	400	...
Naval forces				
Aircraft carriers	10	...	...	...
Cruisers, destroyers, frigates	291	...	210	...
Ocean-going amphibious ships	44	...	19	...
Long-range attack submarines	67	...	142	...

Source: Department of Defense, *Soviet Military Power*, pp. 8, 29, 36, 77, 88, 111.
a. French and Spanish forces are not included in these totals.
b. 1985 figure.

after defending his bridge, numbers are not always everything, a conclusion that Caspar W. Weinberger also reached, at least temporarily, when he wrote:

> Nuclear weapons systems will not be funded merely to make our forces mirror Soviet forces according to some superficial tally of missiles or aircraft deployed in peacetime. Obtaining a facade of symmetry between U.S. and Soviet forces in terms of such simplistic counts is not a requirement for which I would allocate scarce defense dollars.[4]

Several illustrations may help to buttress Weinberger's point. Suppose that Red has 1,500 large ICBMs, each with a ten-megaton warhead that is quite unreliable and inaccurate. Suppose also that the force is based

4. *Department of Defense Annual Report, Fiscal Year 1983*, p. I-17.

in a concentrated and soft configuration. Is that a better force than Blue's 500 ICBMs, with multiple warheads of much lower yield but higher reliability and pinpoint accuracy, especially if Blue's force is dispersed and based in very hard silos? If numbers of ICBMs or even megatons were the sole measure of military power, Red would have to be considered superior to Blue, as the president argues. But given all the other factors in Blue's favor—factors that can be quantified but do not lend themselves easily to simple comparison—a straightforward analysis shows that Blue could knock out Red in a first strike and could still destroy him, even assuming that Red had struck first.

In a comparison closer to reality, it has been widely advertised that the U.S. Navy, having dwindled to 479 so-called battle force ships, had lost its superiority to a Soviet navy consisting of more than 1,000 ships. This alleged loss of superiority was then used as a major justification for building a 600-ship U.S. battle force, although it was never explained why the addition of 121 ships would change American inferiority to American superiority. What the original comparison failed to point out, however, is that the Soviet navy is divided into four fleets, two of which are unlikely to pose much of a threat to U.S. sea lines of communication or to U.S. naval forces. Nor did the comparison mention that the two most threatening fleets—the Northern located in the vicinity of Murmansk and the Pacific based on Vladivostok—would have to undertake hazardous voyages and suffer severe losses trying to reach positions from which they could threaten sea lanes crucial to the United States and its allies. Nothing, finally, was said about the performance of Soviet warships, about the threat to U.S. forces of Russian vessels in the Caspian Sea, or about the inclusion on the Soviet, but not on the U.S., side of small craft incapable of operations on the high seas. In the circumstances, statements about American naval inferiority were bound to be worse than meaningless.

The more sophisticated counts and comparisons of beans will admittedly list other factors such as the performance of weapons, their mobility and vulnerability, the geography in which they might operate, and their logistic support. But only the Army has developed a methodology, and a most troubling one at that, for converting all these variables into scores that lend themselves to comparison.[5] And even these scores, which lead to statements about the relative combat power of Soviet and U.S.

5. For a description of these scores, see William P. Mako, *U.S. Ground Forces and the Defense of Central Europe* (Brookings, 1983), pp. 105–25.

divisions, do not permit inferences by outsiders about how the forces of the one side or the other might be modified. Only the initiated are said to be able to reach the appropriate conclusions.

Military Experience and Judgment

A broader interpretation of this view is that whatever the indicators being used and however they may be used—whether for force assessment or force planning—uniformed personnel (preferably officers) with military experience and judgment are required to interpret the results of any comparison and decide what is to be done about them. There are, however, several reasons why military experience and judgment alone may not suffice to ensure either good assessments or sensible force plans. The most obvious one is that current military experience and judgment do not extend to the conduct of nuclear war. So far, no one has fought such a war, yet it is one of the tragic conditions of U.S. security that nuclear capabilities must be deployed and that judgments must be made about how best to design these capabilities for the deterrence of nuclear and possibly other types of conflict. Military experience and judgment undoubtedly have a role to play even in this strange arena. But here as elsewhere policymakers are entitled to more than assertions and conclusions. Logic and evidence are needed as well.

This requirement is equally great in the assessment and planning of conventional forces. Although the ground of nonnuclear warfare is well trodden, and a number of the U.S. military have fought in nonnuclear campaigns, the ability to command does not automatically translate into the ability to do force planning. The latter demands clear statements of cause and effect and the talent to relate such statements to costs and to provide a sense of how increases and decreases in costs and forces will change the result. Planners must also be able to tell policymakers where one type of capability can substitute for another and why some combinations of these capabilities may be more efficient than others. In principle, knowledgeable military men and women are better qualified to offer these kinds of insights and assessments than anyone else. Currently, however, a number of obstacles stand in the way of their doing so. The biggest rewards go to the officer who shows an aptitude for command and versatility; the specialized skills needed for staff work, and force planning in particular, are not encouraged or even given serious recognition. To the extent that the art of planning is taught, the student is exhorted to describe the threat, formulate the objective, and state the requirement needed to reach the goal. If costs are considered, they enter only after the requirement has been established. As for marginal utility, it rarely affects the solution at all.

None of this should be taken to mean that the military has made no effort to go beyond intuition and rules of thumb in its assessments and planning. Over the years members of the services have developed a number of more systematic techniques for determining force size and composition. They have also taken the lead in encouraging studies at a number of research organizations that would assist and improve their efforts. The Joint Chiefs of Staff already provide the secretary of defense with an annual description of what they call the minimum-risk force, and some of the methodology can be seen at work there. However, the description never explicitly states the odds of success and failure associated with the force, the secretary is never told how the odds might change as a result of variations in the size or composition of the force, and a calculation of alternative defense budgets is the exception rather than the rule. If the secretary were to take literally what he is being told, he would almost have to believe that the minimum-risk force as it emerges from the Joint Staff means success and anything less means disaster. Mr. Micawber might have had something like the minimum-risk force in mind when he said: "Annual income twenty pounds, annual expenditure nineteen nineteen six, result happiness. Annual income twenty pounds, annual expenditures twenty pounds ought and six, result misery."

CHAPTER FIVE

NET ASSESSMENTS AND FORCE PLANNING

STATEMENTS of requirements and measures such as percent of GNP, comparative expenditures, and crude force comparisons are clearly inadequate for force planning. An alternative is needed. Fortunately, it can be found in an adaptation of the force assessment and planning methods developed and sponsored by the military itself. In essence, such an alternative entails the construction of several different forces and their comparison according to the criteria of cost and performance.

Forces, Costs, and Performance

The U.S. military, despite a propensity to argue its case in simple terms and a preference for giving its civilian masters no choice in force size and composition, recognizes that there is no single correct answer to the question of how much is enough. The military also knows that among the many possible answers to that question, some are more efficient than others.

To deal with the issues of sufficiency and efficiency, it is desirable—at least in principle—to compare a substantial number of forces according to the criteria of cost and effectiveness. As a practical matter, the comparisons can be less comprehensive. This is because any new administration inescapably inherits a force and a budget that it may not like but must live with and modify to its own preferences over time. At a minimum, therefore, it is always possible to compare the inherited (or baseline) force with the administration (or programmed) force to see how they differ in cost and performance. It is also feasible to construct and cost an ideal force designed to remedy the weaknesses found in the baseline force, and to do so without the inhibitions imposed by service, congressional, or other parochial preferences. The ideal or combat force, in turn, can be compared with the baseline and programmed forces. Such a comparison should permit insight into relative efficiencies and the principal reasons why they differ.

Tests of Performance

To make these comparisons in a way that emphasizes results rather than a mechanical weighing of capabilities, several tests have to be devised. They should reflect both what the United States wants to be able to accomplish with its armed forces and the conditions that might reasonably obtain in the foreseeable future. However tedious, variables such as measures of effectiveness, the time of the tests, the capabilities of enemies and allies, and military missions have to be defined.

As the State Department might put it, the main military objective of the United States is to deter attacks on the nation's vital interests. However, the only known way to measure the effectiveness of a deterrent, nuclear or conventional, is paradoxically to hypothesize a conflict in which enemy forces are pitted against those of the United States and, where relevant, its allies at a particular time and in a particular setting. The assumption is that if U.S. forces are estimated to have a significant probability of achieving their objectives, they will constitute a powerful (but not necessarily perfect) deterrent to hostile military action under comparable or less demanding conditions.

One challenge of this approach is that no one can foresee with confidence where, when, or how the nuclear or conventional capabilities of the United States might be tested, whether by the threat of military force or by its use. There are, however, certain contingencies that have come to be used as the most rigorous (and realistic) tests that a particular capability could be asked to undergo. Thus, the standard test for the strategic nuclear forces is their ability to retaliate in the wake of a well-executed Soviet surprise attack, even though other and less demanding nuclear attacks may be more plausible.

Although the United States asserts as declaratory policy that, if necessary, it would make a first use of nuclear weapons in a tactical campaign, force planners can have no assurance that the president would or even could take this initiative. Accordingly, the theater or tactical nuclear capabilities are best assessed on the assumption of a Soviet first strike in Europe aimed at destroying U.S. and allied nuclear assets, followed by U.S. retaliation with its surviving forces. However, it is the exception rather than the rule that these capabilities undergo any military assessment at all.

The most demanding task faced by the conventional land and tactical air forces of the United States is—in conjunction with allied capabilities—both to stop an attack on Central Europe by the Warsaw Pact and to prevent the Soviet Union from conquering Norway and gaining greater operating room for its northern fleet. Conceivably, more or less simultaneously, U.S. ground and tactical air forces would be called upon to prevent a Soviet march through Iran aimed at seizing the oil fields in the vicinity of the Persian Gulf and to reinforce existing U.S. deployments in South Korea.

It has become customary, because of the assumption of simultaneity, to describe U.S. conventional strategy as the ability, in conjunction with allies, to fight two and one-half or one and one-half wars. Although, rhetorically, the Reagan administration has rejected this strategy in favor of being prepared to fight a worldwide conventional war (as though it would not be conducted in particular theaters), this administration too has obviously used a limited number of contingencies as the basis for testing the adequacy of its programs. But however the force planning strategy may be labeled, it tends to be a misnomer in that it omits at least three critical tasks and tests: the ability of U.S. long-range airlift and sealift to deploy overseas the land and tactical air forces necessary for the defense of key theaters; the ability of naval and Marine Corps forces both to keep open the sea lines of communication for the resupply of the forces overseas and to prevent Soviet naval forces from breaking out of their constricting geographical confines; and the ability of the conventional forces to sustain themselves in combat with munitions and other consumables. In other words, current force planning concepts not only call for the forces to deal with a range of nuclear and conventional contingencies; they also require the United States to get the designated forces to the threatened theaters in a timely fashion, to keep open the lines of communication to them, and to sustain their operations for at least as long as their enemies can.

Two critical assumptions underlie this kind of planning and assessment. First, while the planning contingencies may never materialize, their importance is such that the United States and its allies should have the forces necessary to deal with (and preferably deter) them. Second, if the forces are available for the planning contingencies, they will probably be sufficient for any other contingencies that might arise, as indeed they might, say, in Latin America, Africa, or the Middle East. Thus, the performance of the baseline, programmed, and combat forces in responding to these planning contingencies can be considered a reasonable if imperfect test of their ability to uphold the main conditions of U.S. security.

Scheduling the Tests

A great deal more than the choice of contingencies must go into the establishment of these tests of performance. One crucial decision concerns the time at which the planning contingencies are assumed to occur. Timing is important because it will determine considerably the characteristics of the forces committed to the contingencies. The baseline force warrants being tested twice: once in 1981 when the Reagan administration inherited the force, and once at a later date when the administration's programs have taken full effect. The test in 1981 is needed primarily as a benchmark from which to measure any subsequent changes in performance. But the test also permits the programmed and combat forces to be compared against this benchmark as well as with each other.

The programmed force currently differs so little from the baseline force because the changes sought by the Defense Department's original program in 1982 will not be fully realized until about 1992. By then, however, the programmed force will presumably be at least qualitatively different and will have undergone the resuscitation deemed vital by the president and the secretary of defense. It seems appropriate, therefore, to use 1992 as the year in which all three forces are tested.

The need, in fairness to the Reagan administration's plans, to conduct the comparisons so many years in the future inevitably raises the question of how much Soviet and other military capabilities, including those of U.S. allies, will have changed during the interval. Although no certain answer can be given, these capabilities can be projected under various assumptions about budget constraints and resource allocations.

Enemies

For purposes of force planning, the capabilities of the Soviet Union and its satellites have to be considered the principal military threat to U.S. interests and the forces against which it is most appropriate to design and test U.S. forces. Presumably, if the United States and its allies can cope successfully with these forces, they can also deal with less difficult contingencies and with peacetime demands for deployments and demonstrations, provided, of course, that the necessary matériel and training for these lesser cases can be made available as needed.

At the heart of the potential threat is the military machine of the USSR. Despite the many difficulties that have afflicted that machine during the past decade, despite the exaggeration of its technological progress and numerical superiority, and despite its mediocre performance in Afghanistan and the ability of the Israelis to destroy its equipment in the Bekaa Valley of Lebanon, there should be no complacency about its future effectiveness.

Arms control agreements may be able to affect this performance on the margin. They may succeed in making estimates of future strategic nuclear capabilities more reliable and in reducing the arenas of competition and the incentives for further buildups. But it is unlikely that, by themselves, agreements will forestall the USSR from improving its nuclear and conventional forces during the next seven years.

How large these improvements will be, and where they will occur, are two of the largest uncertainties with which force planning and assessment must contend. They are constant invitations to pessimistic planning. However, cost, technology, and time are constraints that Soviet leaders, like other mortals, must face. As far as cost is concerned, it seems reasonable to believe, at worst, that the Soviet military budget will resume its real growth of about 3 to 4 percent a year. Over a five-year period that would provide the Soviet military with a cumulative real increase of about $100 billion (U.S. prices) for the further expansion, modernization, and additional operating costs of its forces. Although that is a good deal of money, it would be quickly eaten up by the gradual modernization of the existing force structure with more sophisticated weapons and the higher operating and support costs that would accompany their deployment.

To what extent might improvements be expected by 1992? To test the U.S. baseline, programmed, and combat forces, several assumptions have been made based on directions in which the USSR is currently moving.

Soviet strategic nuclear capabilities are strengthened in two main areas. The offense receives 660 mobile ICBMs and the SS-18 and SS-19 missiles are upgraded to the point where each U.S. silo has only a 4 percent probability of survival. Despite these changes, the U.S. list of

Table 5-1. Estimated Soviet Nuclear Capabilities, 1992

Nuclear forces	*Launchers*	*Warheads*
Strategic[a]		
Intercontinental ballistic missiles (ICBMs)	1,398	8,408[b]
Submarine-launched ballistic missiles (SLBMs)	871	3,951
Bombers	300	2,400
Total	2,569	14,759
Theater		
SLBMs	39[c]	39
Mid-range ballistic missiles (MRBMs)	387	1,161
Sea-launched cruise missiles (SLCMs)	645	645
Medium-range bombers	250[d]	1,000
Short-range bombers	980	980
Short-range missiles	835	835
Artillery tubes	900	900
Total	4,036	5,560

Sources: *FAS Public Interest Report,* vol. 37 (October 1984), p. 12; Department of Defense, *Soviet Military Power* (Government Printing Office, 1985), pp. 36–39; and author's estimates.

a. It is assumed that SALT constraints continue to be accepted in the planning of the strategic forces.

b. At best 5,380 of these warheads are considered to have a high probability of kill (about 80 percent) against hard targets.

c. These consist of relatively old Golf submarines. In addition, the USSR may still deploy variable-range ICBMs that could be used against targets on the periphery of the Soviet Union.

d. These do not include Backfire bombers, which (to follow the current fashion, however mistaken) are counted as part of the strategic nuclear forces.

targets in the Soviet Union remains constant. This happens because, as the Russians add mobile ICBMs, they dismantle an equivalent number of their older silos. On the defensive side, they upgrade further the Moscow antiballistic missile defense system and obtain a 50 percent kill probability against the first one hundred warheads that enter its area of interception. Soviet antisubmarine warfare capability also improves: it has a 10 percent probability of kill against U.S. on-station submarines equipped with the older and shorter-range Poseidon ballistic missiles. Perhaps most important, Russian air defenses are modernized at last with the all-azimuth deployment of a look-down, shoot-down capability that is equivalent to the first generation of U.S. airborne warning and control systems and F-15 fighters in its ability to track and engage low-altitude attackers. This modernization seriously affects the probability that bombers, including the B-1B, can reach their targets, but reduces only minimally the penetration probability of U.S. cruise missiles. Estimated Soviet nuclear capabilities in 1992 are summarized in table 5-1.

Soviet and satellite conventional forces do not change to any marked degree in size and firepower, but they do improve in kill probabilities by nearly 13 percent, or the amount equivalent to the increment of resources

Table 5-2. Number and Readiness of Warsaw Pact Divisions

	Number and readiness of divisions															
	M + 4[a]				*M + 9*[a]				*M + 14*[a]				*M + 90*[a]			
Country	*I*[b]	*II*[c]	*III*[d]	*Total*	*I*[b]	*II*[c]	*III*[d]	*Total*	*I*[b]	*II*[c]	*III*[d]	*Total*	*I*[b]	*II*[c]	*III*[d]	*Total*
Soviet Union	19	...	...	19	26	...	...	26	30	4	25	59	30	4	45	79
East Germany	6	...	...	6	6	...	...	6	6	...	...	6	6	...	...	6
Czechoslovakia	5	...	...	5	4	3	3	10	4	3	3	10	4	3	3	10
Poland	...	...	...	...	10	...	5	15	10	...	5	15	10	...	5	15
Total	30	...	...	30	46	3	8	57	50	7	33	90	50	7	53	110

Sources: William W. Kaufmann, "Nonnuclear Deterrence," in John D. Steinbruner and Leon V. Sigal, eds., *Alliance Security: NATO and the No-First-Use Question* (Brookings, 1983), p. 60; and author's estimates.

a. M stands for mobilization day.

b. Divisions are considered combat-ready: 75–100 percent of personnel; complete set of modern equipment.

c. Divisions are at reduced strength: 50–75 percent of personnel; complete set of combat vehicles.

d. Divisions are at cadre strength: less than 50 percent of personnel; complete set of combat vehicles (many of them obsolescent).

available for the deployment of upgraded old or more sophisticated new systems. Perhaps this assumption is excessively conservative, since it is rarely true that effectiveness increases commensurately with increases in costs. Furthermore, no firm evidence suggests that the Russians can do better than the United States in this respect. Even less realistically, the satellite forces of the Soviet Union are allowed equivalent increases in kill probabilities, although it is doubtful that the satellites' defense budgets will rise as rapidly as that of the Soviet Union.

Despite these improvements, all Warsaw Pact ground forces (including Soviet ground forces located in southern and far eastern military districts) continue to keep most of their units at low states of combat readiness, as shown in table 5-2, primarily because these units depend on the call-up of reserve personnel to reach full strength. The reservists, for the most part, receive little or no training once they have completed their conscripted period of service. What are categorized as category II and III divisions can probably be mobilized and deployed as rapidly as the combat-ready category I divisions. But when they are included in the surprise attacks favored by the U.S. military (which, depending on their scale, are launched after four, nine, or fourteen days of mobilization and deployment, designated as M + 4, M + 9, M + 14), the divisions' effectiveness is severely downgraded even though they are credited, conservatively, with the full firepower and mobility of which they are capable. Only after ninety days of mobilization, deployment, and training in the field are they rated at full effectiveness. Even with these limitations and the problems that plague the Soviet transportation system, the USSR is able by 1992 to initiate and sustain more or less simultaneous attacks in such widely separated theaters as north Norway, Central Europe, and Iran. And Soviet frontal aviation has the capability by 1992 to conduct air superiority and interdiction as well as close air support campaigns in all three theaters, while air defenses in Eastern Europe are dense, well coordinated, and capable of intercepting U.S. and allied aircraft as they enter and leave the airspace of the Warsaw Pact. At sea, all Soviet attack submarines are committed to attacks on allied sea lines of communication to the engaged theaters even though Soviet naval doctrine appears to assign them first and foremost to the protection of the Russian ballistic missile submarines. Naval Backfire bombers are also committed against allied convoys despite the threat of U.S. carrier-based attacks on the ports of the Northern and Pacific fleets. Furthermore, the Soviet navy has established two heavily protected overseas main bases from which

Table 5-3. U.S. and Allied Forces Available for the Defense of Central Europe

	Warsaw Pact attack									
	M + 4		M + 9		M + 14		M + 90[a]		M + 90[b]	
Country	*Divisions*	*Combat aircraft*	*Divisions*	*Combat aircraft*	*Divisions*	*Combat aircraft*	*Divisions*	*Combat aircraft*	*Divisions*	*Combat aircraft*
Belgium	2	126	2⅔	126	2⅔	126	2⅔	126	3⅓	126
Britain	4	340	5	340	5	340	5	340	6	340
Canada	⅓	42	⅔	66	⅔	66	⅔	66	⅔	66
Denmark	1	100	1⅔	100	1⅔	100	1⅔	100	1⅔	100
France	3	405	7	405	7	405	7	405	10	405
Netherlands	2⅓	126	3	126	3	126	3	126	3⅔	126
West Germany	12	390	14	390	14	390	14	390	14⅓	390
Subtotal	24⅔	1,529	34	1,553	34	1,553	34	1,553	39⅔	1,553
United States	5⅔	612	9	1,440	12	1,872	15	1,872	23	2,520
Total	30⅓	2,141	43	2,993	46	3,425	49	3,425	62⅔	4,073

Sources: Kaufmann, "Nonnuclear Deterrence," pp. 62, 77; and author's estimates.

a. These are the forces that would be available at M+90 under current assumptions about U.S. and allied mobilization and deployment.

b. These are the forces that would be available at M+90 if the allies increased modestly their investment in reserve ground forces and the United States adopted the combat force, with its emphasis on more ready reserve divisions and more close air support aircraft.

they can support the operation of attack submarines and surface combatants.

Allies

When the USSR launches a surprise strategic nuclear attack, only U.S. forces are counted as far as retaliation and target coverage are concerned, even though British, French, and Chinese strategic capabilities might also be included in the Soviet attack. In conventional contingencies, however, allied forces play a major role (except in the Persian Gulf) and U.S. ground combat and tactical air forces in Europe and Korea supplement them as necessary to maintain a forward defense. It is assumed that the allies improve their conventional capabilities as much as the USSR and its satellites do, so that the gap in performance between them neither widens nor narrows in the relevant theaters. Thus, in Central Europe the allies continue to provide up to about forty divisions (including French as well as British, Canadian, Belgian, Dutch, Danish, and German units) and 1,553 combat aircraft, which will do as well (or badly) against an invasion by the Warsaw Pact in 1992 as they did in 1981 (table 5-3). If this seems an optimistic assumption, at least it emphasizes that any change in overall performance depends on what the United States does to improve its capabilities.

The Reagan administration has emphasized the new directions on which it has embarked in defense. However, for force planning and assessment, the administration's defense objectives—which must also be specified—do not differ materially from those of its predecessors. Consequently, standard goals and measures of effectiveness are used in evaluating the baseline, programmed, and combat forces. Note, though, that since a statement of probability usually provides the most appropriate measure of effectiveness in completing a mission, what constitutes sufficiency is a matter of judgment, especially since effectiveness may not increase commensurately with the increase in cost. Is it better to have a 60 or a 70 percent probability of destroying 5,000 enemy targets if the 10 percent increase in probability is accompanied by a 20 percent increase in cost?

Strategic Nuclear Retaliation

The question becomes particularly relevant in the evaluation of strategic nuclear forces. Although General Douglas MacArthur proclaimed that in war there is no substitute for victory, and although current Pentagon talk is once again of prevailing, there is no sure way of winning a strategic nuclear exchange other than by eliminating the other side's nuclear weapons with a disarming attack. Under present and foreseeable conditions, however, that does not appear to be a realistic objective. Other ways of gaining meaningful leverage with these forces have been invented, but none of them promises much confidence of success and all of them risk unacceptable damage in the process. For these reasons the most sensible objective for the U.S. strategic nuclear forces continues to be the denial to the USSR of a successful disarming first strike and any more protracted attempt to gain an exploitable military advantage.

What this objective signifies operationally, since the strategic forces can only blow things up, is the ability to cover such a comprehensive list of targets on a second strike that no matter what strategy an enemy might adopt, it could be stalemated by the U.S. retaliation—whether against his remaining forces or against other vital assets. Several target lists can obviously be compiled for this purpose. But a rigorous test of the ability of the strategic nuclear forces to execute this countervailing strategy would be set by a list of 5,274 separate aiming points. Such a list would be made up of hard and soft military targets (containing both nuclear and conventional forces), economic and logistical assets not collocated with cities, and various aiming points in urban-industrial areas.

In the event of a nuclear attack the national command authority would have a number of options available, ranging from limited retaliation against only a small portion of the total target list—while withholding the rest of the force—to the release of the entire surviving force and attacks on the full list of targets. For force planning and assessment, however, the most appropriate measure of force performance is the number and mix of targets that can be destroyed without retargeting.

This last constraint is imposed for several reasons. Although the strategic nuclear forces can be retargeted after an enemy attack, the process is fairly time-consuming and could have, in any event, been damaged by the attack. Furthermore, there could well be great uncertainty about the number and composition of the surviving U.S. capability. Because of these limitations, the strategic forces are acquired on the assumption that no retargeting will occur and that each aiming point in the Single Integrated Operational Plan must have one or more warheads dedicated to it. Not only does such a policy entail larger forces than would be necessary if retargeting could be counted on; it virtually ensures that the targets assigned to the forces prior to the enemy first strike are the ones that will be attacked.

Obviously, longer or shorter target lists could be chosen. If the list were longer, all three forces (baseline, programmed, and combat) would do less well. Relatively, though, their performances would remain about the same. If the list were much shorter and more selective—if, for example, most hard targets were omitted—differences in performance among the three forces would be negligible and it could be argued that all of them are too large. All of which illustrates how sensitive the performance and hence the size and composition of the force are to the decision about the objective.

Tactical Nuclear Retaliation

Nowhere is this consideration more evident than in cases involving tactical nuclear capabilities. Although the United States deploys these weapons in many locations, including ships and submarines at sea, the main theater of concern for force planning and assessment is Central Europe. And the main objective there is to prevent a successful invasion by the Warsaw Pact. The three services, to this end, continue to acquire nuclear weapons and launchers for them on the premise that a tactical exchange of nuclear weapons would resemble a conventional engagement and that, to halt the invasion, they would have to fight a relatively protracted campaign. It is worth remembering, however, that conventional forces engage in maneuvers, advances, and retreats, and consume large stocks of munitions over fairly long periods precisely because high explosives (and chemical weapons for that matter) are not very efficient. However, these same forces and their logistical support can be rapidly devastated by a relatively small number of nuclear weapons. Thus, rather than continue the pretense that theater nuclear warfare would be just like its nonnuclear counterpart, only bloodier, it makes sense to choose for the tactical nuclear forces the same kinds of objectives set for the strategic nuclear forces. This means the establishment of a target list, the destruction of which would give a high probability of halting an invasion of Western Europe by the Warsaw Pact. The ability on a second strike (since the USSR might well preempt a U.S. first use) to knock out some 1,500 targets, including enemy troops at the front, should be more than enough to cause the collapse of such an invasion in short order. Whether the collapse would be accompanied by the wreckage of Eastern and Western Europe and the rapid escalation of the conflict to a strategic nuclear exchange are two uncertainties that make any use of nuclear weapons look like an appalling gamble to leaders of the West, whatever their rhetoric, and probably to leaders of the East as well.

Conventional Forces

There is probably less controversy about the objectives for the conventional forces, and how to measure the performance of these forces, than there is about the feasibility of reaching their goals. Because of longstanding U.S. commitments it has to be assumed for purposes of force planning and assessment that at least the initial objective, whether in Central Europe, north of Saudi Arabia, or in South Korea, will be to maintain a defense sufficiently far forward to prevent the loss of allied territory to the attacker.

Ground and Tactical Air Campaigns

Such a defense usually relies on the performance of three missions: the conduct of the direct response to the attack with a combination of ground forces and close air support aircraft; the acquisition of air superiority by fighter aircraft and air defenses; and the initiation of an interdiction campaign by attack aircraft to limit reinforcements and supplies to enemy forces in and near the forward edge of the battle, or front. The key to the success of this kind of campaign is the ability of the ground forces and close air support aircraft to prevent a major penetration or breakdown of the front for as long as the enemy can sustain an offensive.

It is possible to measure the probability that such a result can be achieved and to translate that probability into expected outcomes, the time it would take to achieve them, and the losses that would be suffered by both sides. Furthermore, since close air support can substitute within limits for ground forces, the utility of the substitution can be analyzed. Conceivably, the air superiority and interdiction missions, by preventing the enemy from disrupting U.S. and allied lines of communication and by reducing enemy reinforcements and supplies, can significantly affect the probability of sustaining a forward defense. However, since it is often assumed that the initial attacks would be intense and that the outcome of a modern conventional campaign would be rapidly decided—within a few days or weeks—the air superiority and interdiction efforts must deliver on their promise of protecting allied logistics and seriously damaging enemy logistics within a few days. The success in restricting the flow of tonnage below the level needed by the enemy to sustain his offensive can be measured (as can the enemy's success in restricting U.S. and allied tonnages). Based on those results and other tests, judgments can also be made about the optimum allocation of resources to the three major missions.

Strategic Mobility

As has been stressed for many years, it makes sense to maintain some U.S. ground and tactical air forces in sensitive overseas regions such as Central Europe and South Korea. They serve both to symbolize the U.S. determination to defend these areas and to guard against attacks that could occur with little or no warning. A further hedge against short-warning attacks can be acquired by prepositioning weapons, equipment, and supplies in or near an important theater—on land (as in Europe) or at sea (in the Indian Ocean, but within a few steaming days by ship to the head of the Persian Gulf). But both these measures are expensive, and once the matériel and forces have been put in place, it becomes difficult either to withdraw them or to use them elsewhere, even temporarily. Furthermore, because of the worldwide nature of U.S. interests and uncertainty about where the next crisis might arise or where the president might decide to commit forces, it has seemed wise to keep a large percentage of U.S. ground and tactical air forces in the continental United States, and to have the capability to deploy those forces rapidly to wherever they might be ordered.

In the 1960s an attempt was made to achieve a rapid deployment capability consisting of three elements: prepositioned matériel to which a small number of U.S. divisions could be deployed, if necessary, by civilian aircraft; widebodied airlift aircraft for the rapid movement of both the people and the equipment for a few more divisions; and large, fast ships loaded with equipment and supplies that would be stationed near critical theaters in peacetime and could steam rapidly to key ports if a crisis occurred. The effort was only partly successful, however, and emphasis since then has been put primarily on prepositioning and widebodied airlift.

To some extent this shift has resulted from a parochial congressional reluctance to support a big investment in fast sealift that might be used in peacetime to compete with the heavily subsidized U.S. Merchant Marine. But a more important factor has been the change in assumption about how quickly enemy forces could prepare for attacks in the key force planning theaters. Generally speaking these kinds of assumptions should not break any physical laws or exceed the limits of plausibility. Whether the USSR and its satellites could mobilize and deploy large forces in the short intervals currently assumed, or whether they would do so without the preliminaries of a major crisis, can be argued. What is not at all plausible is that as many as ninety divisions could be deployed in fourteen days, and that all the forces engaged in the deployment could be considered fully combat-ready. For various reasons, including the fact that most divisions in the Warsaw Pact inventory are about the equivalent of low-readiness U.S. National Guard and Reserve units, either the deployment can be rapid—with many of the divisions at low levels of effectiveness—or the preparations can be a great deal slower—with all the forces assumed to be fully effective. But force planners cannot have it both ways.

If the deployment is rapid, airlift is the only serious recourse for the United States. But the pressure on it will be less than usually assumed because fewer U.S. forces will be needed to counter the attack. If the deployment is relatively slow—taking months rather than days—more U.S. forces will have to be deployed, but they can travel by fast sealift rather than airlift. What is more, while it has been assumed that rapid mobilization and deployment by the USSR and its satellites would constitute the worst threat, both a more probable and a more difficult attack would eventuate from a fairly slow enemy mobilization and the deployment of a larger number of units than is often assumed. Thus, it seems appropriate to test the ability of the strategic mobility capability against both fast and slow enemy deployments, but to set the objectives (measured in tons needed in a given period) somewhat lower than customary for the short-warning cases, and at least 20 percent higher in the long-warning cases, which might be about ninety days in theaters such as Central Europe and the Persian Gulf.

Naval Forces

The U.S. Navy is in some sense a microcosm of the Defense Department: it has its own army (the Marine Corps), navy, and air force (carrier-based fighters and land-based patrol aircraft). Consequently, the Navy can not only engage enemy naval forces, but also duplicate with its amphibious forces and carrier battle groups some of the functions performed by the Army and the Air Force. There is no doubt, however, that a substantial premium must be paid for ground and tactical air forces based on platforms that move rapidly and that have to be heavily protected by aircraft, missiles, and guns that are also on mobile platforms. Where land-based ground and tactical air forces are free to operate, they always have a large cost advantage over their naval counterparts. But as was exemplified by the recent war over the Falkland Islands, there are parts of the world the United States cannot readily reach, at least initially, with land-based forces. In principle, therefore, a role exists for carrier battle groups and amphibious forces; whether, in practice, clear-cut missions can be found for them is in dispute.

Sea Lane Protection

It has to be assumed for planning purposes that in the event of one or more conventional conflicts involving the USSR, the Navy would have the mission, in conjunction with allies, of keeping open the sea lanes from North America to Europe, the Persian Gulf, and Japan and Korea—assuming that French, Italian, and Spanish naval forces by themselves would be adequate to patrol the sea lanes in the Mediterranean. This mission would call primarily on the Navy's mine warfare forces, attack submarines, patrol aircraft, and surface combatants, both to block the main passages that Soviet forces would have to transit to attack U.S. and allied shipping and to convoy economic and military supplies to their wartime destinations.

Power Projection

The amphibious forces would have no direct role to play in these operations. Furthermore, it is by no means clear that the carrier battle groups—considering the availability of land bases in Newfoundland, Britain, Japan, and the Philippines from which long-range Soviet aircraft could be intercepted—could be justified solely for this purpose, especially given their cost. However, if the USSR could break out of its current geographical confinement in Murmansk and Vladivostok and establish home ports overseas, from which it could operate without crossing the antisubmarine warfare barriers that the United States can now construct, the cost of protecting the key sea lines of communication would rise substantially.

The USSR seems determined to make this possibility a reality in its activities in the Atlantic, the Red Sea, and Vietnam, areas that could be bombed but not seized by U.S. land-based forces. Amphibious forces and carrier battle groups are better suited to this mission. Furthermore, they could be essential to effecting an initial lodgment in a remote area such as the Persian Gulf.

Whether from conviction or from the simple desire to justify three additional carrier battle groups, the Navy has also asserted a requirement for the capability to attack the Soviet fleets in their home ports of Murmansk and Vladivostok. The most recent assertion of this requirement has been based on the argument that if the USSR should attack in an area of U.S. vulnerability, the United States in response should be able to escalate the conflict horizontally by using carrier battle groups to strike at a point of Soviet vulnerability. The more enduring justification for such a power projection force rests on the traditional naval axiom that it is better to bring the enemy fleet to action and destroy it than to engage in the more defensive, wearing, and mundane functions of long-range blockade and convoy duty. However, it is also true that the Navy wants the capability both to do a Trafalgar or a Copenhagen on the Soviet fleets in the best tradition of Lord Nelson and to perform the tasks of sea control, however unheroic they may be.

In a short war sea control would not serve much of a purpose because prepositioned matériel and supplies overseas would presumably last long enough to support deployed forces for at least a month and because the conflict would probably have ended before the first convoys could arrive. In a war lasting more than a month, however, additional tonnage equal to daily rates of consumption of the overseas forces would have to be delivered, and delivered before the prepositioned stocks had been exhausted. The percent of needed tonnage that is deliverable is the simplest way to measure the effectiveness of the sea control forces.

Though there may be several ways to neutralize Soviet overseas bases, the surest if not the simplest method is to capture them, especially if they contain hardened facilities such as submarine pens. On the conservative assumption that two overseas bases might have to be attacked at the outset of a conflict and that, simultaneously, carrier battle groups and amphibious forces would be needed to obtain a beachhead in the Persian Gulf area, it is possible to use the probability of destroying the forces protecting these locations—presumably with Marine forces and naval fighter-attack aircraft—as the measure of effectiveness in these operations.

It seems reasonable to assume for planning purposes that the Persian Gulf and any Soviet overseas bases would have first call on the carrier battle groups in the event of a major conflict. Hence any attack on Murmansk or Vladivostok either would have to be made with residual carrier assets (always supposing that only U.S. naval forces can attack Soviet naval forces) or would have to await the release of other carriers from their initial missions. On the premise that the doctrine of command of the seas and the rhetoric of horizontal escalation would inspire an early attack, the test of performance is made with whatever carrier battle groups are not tied down elsewhere, and two measures of effectiveness can be used. The first estimates the overall probability of destroying the Soviet fleet in and around the port—assuming optimistically that it has not dispersed upon warning of the attack, warning that might be received several days in advance. Since Soviet long-range attack submarines will continue to be the main threat to allied shipping, the second measure estimates the expected survival of the various components of the Soviet fleet and of the U.S. carrier battle groups. In this connection the U.S. attack does not distinguish between ballistic missile and attack submarines.

The Three Forces

The main elements of the three forces put to these tests are summarized in table 5-4. Although, with one exception, all the tests of effectiveness are conducted under what are projected to be the conditions existing in 1992, the costs of the forces are calculated only through fiscal 1990, the end of the current five-year planning period. This cutoff was chosen on the assumption that funding after 1990 would not have a major impact on the size and composition or performance of the forces in 1992.

The Baseline Force

Both the baseline and the combat forces are artificial constructs. The baseline force is the capability that the Reagan administration inherited

Table 5-4. Salient Baseline, Programmed, and Combat Force Capabilities, Fiscal Year 1992

Capability	*Force*		
	Baseline	*Programmed*	*Combat*
Strategic nuclear warheads[a]	9,592	12,516	11,058
Theater nuclear warheads	9,000	7,600	4,400
Conventional forces deployable by M+90			
Land-based forces			
Army division-equivalents	18	20	33
Air Force tactical fighter squadrons	78	78	122
Power projection forces			
Aircraft carrier battle groups	12	15	12
Marine Corps			
Divisions	4	4	4
Tactical fighter squadrons	33	33	33
Amphibious lift (number of divisions transportable)	1.1	1.5	1.1
Sea lane protection forces			
Antisubmarine warfare barrier capabilities[b]	3	3	3
Convoys escorted per month	7	8	9
Rapid deployment capabilities			
Airlift (millions of ton-miles per day)[c]	32.4	66.0	32.4
Fast deployment ships	8	8	38

Source: Author's estimates. See also table 5-14.
a. On launchers.
b. Sufficient mines, attack submarines, and patrol aircraft to establish barriers across such areas as the Greenland–Iceland–United Kingdom gaps.
c. Only long-range, wide-bodied aircraft are included.

in 1981. Its performance is frozen at the 1981 level, but its costs are calculated in fiscal 1986 dollars (tables 5-5, 5-6, and 5-7). Alone among the three forces, the baseline force is put to the test against the 1981 as well as against the 1992 threat.

The Combat Force

Although the combat force's hypothetical planners accept the dynamism of the military competition with the USSR, they see Soviet programs as more evolutionary than revolutionary, directed more toward the modernization of existing forces and the gradual exploitation of new technologies than toward the rapid buildup of military capabilities that has been the hallmark of some prewar periods. Accordingly, these planners attempt to remove what they consider the main vulnerabilities in the U.S. baseline posture and try to increase its probability of success in reaching U.S. objectives. But they do not press for crash programs.

Table 5-5. Cost of the Three Forces by Major Capability, Fiscal Year 1990
Billions of 1986 dollars

	Budget authority		
Capability	*Baseline force*	*Programmed force*	*Combat force*
Strategic nuclear forces	29.3	59.5	41.8
Theater nuclear forces	2.4	4.9	3.5
Land forces	52.3	97.8	80.5
Land-based tactical air forces	34.4	57.5	53.4
Navy tactical air and antisubmarine warfare forces	14.1	39.2	18.8
Navy ships and submarines	35.4	78.1	57.3
Airlift and sealift	6.3	12.3	10.9
Intelligence and communications	22.0	38.5	31.2
Total[a]	196.2	387.8	297.3

Sources: *Department of Defense Annual Report, Fiscal Year 1982*, p. C-2; *Historical Tables, Budget of the United States Government, Fiscal Year 1986*, p. 5-1(3); Department of Defense, Office of the Assistant Secretary of Defense (Comptroller), *National Defense Budget Estimates for FY 1986*, p. 48; and author's estimates.
a. Does not include military retired pay accrual.

Table 5-6. Cost of the Three Forces by Major Planning Contingency, Fiscal Year 1990
Billions of 1986 dollars

	Budget authority		
Planning contingency[a]	*Baseline force*	*Programmed force*	*Combat force*
Strategic nuclear retaliation	29.3	59.5	41.8
Theater nuclear retaliation	2.4	4.9	3.5
Conventional defense of:			
Central Europe	59.7	97.6	91.8
North Norway	12.0	22.8	17.8
Mediterranean states	. . .	9.1	. . .
Atlantic and Caribbean	12.3	27.2	18.8
Persian Gulf	20.0	40.8	43.7
Korea	6.3	14.1	11.7
Pacific and Indian Oceans	15.8	33.6	25.0
CONUS,[b] Alaska, Panama	16.4	39.7	12.0
Intelligence and communications	22.0	38.5	31.2
Total[c]	196.2	387.8	297.3

Sources: Table 5-5; and author's estimates.
a. Allocations are based on the assumption that all planning contingencies occur simultaneously. All airlift and sealift costs are allocated to Central Europe and the Persian Gulf.
b. Continental United States. The costs of forces that cannot be deployed by M+90, or that are in training, maintenance, or overhaul are allocated to CONUS.
c. Does not include military retired pay accrual.

Table 5-7. Five-Year Costs of the Three Forces (Retired Pay Excluded), Fiscal Years 1986–90

Billions of 1986 dollars

Item	*1986*	*1987*	*1988*	*1989*	*1990*	*Total*
Budget authority						
Baseline force	196.2	196.2	196.2	196.2	196.2	981.0
Combat force	264.1	272.0	280.2	288.6	297.3	1,402.2
Programmed force	295.9	321.5	351.2	369.4	387.8	1,725.8
Outlays						
Baseline force	176.6	176.6	176.6	176.6	176.6	883.0
Combat force	237.7	244.8	252.2	259.7	267.5	1,261.9
Programmed force	259.7	281.1	301.2	317.6	335.2	1,494.8

Sources: *Department of Defense Annual Report to the Congress, Fiscal Year 1986*, p. 78; tables 5-5 and 5-6; and author's estimates.

Instead, they accept the rule that the life cycle of major weapons platforms is, on the average, about twenty years, and they pace upgrading and modernization programs to accommodate that cycle. In doing so, however, they assume that although previously the United States could meet the uneven challenge from the USSR with budgets that, on the average, remained level in real terms, the gradual improvements made in Soviet capabilities during the last twenty years have finally reached the point where U.S. defense budgets from now on (1981) will have to increase moderately in real terms to ensure the main conditions of U.S. security. Planners also assume, perhaps optimistically, that the principal allies of the United States will make a comparable defense effort. They find that steady modernization of the baseline force (including the National Guard and Reserve forces) and its maintenance at a high level of operational readiness can be accomplished with defense budget authority (not including military retired pay) that increases in real terms by 3 percent a year starting in fiscal 1982. The capabilities available in 1992 reflect these decisions and constraints. Costs and allocations of the combat force are shown in tables 5-5, 5-6, and 5-7.

The Programmed Force

The programmed force is in a sense an arbitrary construct as well, though to a lesser extent than the baseline force and the combat force. This is true because it is the product not only of the decisions made between 1981 and 1985, but also of the five-year defense program for fiscal 1986 through 1990 submitted by the president in February 1985, a plan that has little chance of being fully implemented. Despite this likelihood, the president's program for the rest of the decade (as shown in tables 5-5, 5-6, and 5-7) is used for several reasons. The program is the only administration plan currently available, and it is presumably an accurate reflection of what the Defense Department set out to accomplish in 1981: a ten-year modernization of the baseline force, accompanied by some expansion of the force structure and an increase in its readiness and sustainability. With that emphasis continued, the capabilities that emerge in 1992 not only mirror the administration's ambitions but also show the effects of a planning approach, depending as it does on the preferences of the three services, that differs markedly from the one used to develop the combat force, even though both approaches are assumed to have essentially the same objectives.

Performance of the Forces

A comparison of the performance of the baseline, programmed, and combat forces is summarized in table 5-8. The baseline force probably does less badly in 1992 than the rhetoric about its decrepitude in 1981 would have suggested. For some constituencies, in fact, its performance may be considered quite acceptable. However, when its 1992 performance is compared with what it could do in 1981, the considerable erosion brought on by the steady improvement in Soviet capabilities quickly becomes evident. Whatever the defects in the Reagan administration's assessment of the military situation in 1981 and prior years, both Reagan and his secretary of defense, like Carter and his secretary of defense, had a strong case for making real increases in the defense budget and forestalling a deterioration of this character.

The programmed force clearly reverses the erosion that marks the baseline force. By 1992, in fact, it brings performance back roughly to where it was in 1981, despite the Soviet advances. However, the programmed force still does not provide as much as a fifty-fifty chance of conducting a successful conventional defense in either Central Europe or the Persian Gulf. Furthermore, despite the addition of three carrier battle groups to the U.S. fleet, it has only one chance in five of destroying the Soviet fleet in Murmansk and, when the expected values are calculated, does poorly against the submarines assumed to be in port. In the process, it loses all three of the attacking carriers.

Table 5-8. Summary of the Performance of the Three Forces

		Force			
		Baseline			
Contingency	*Measure of effectiveness*	*1981*	*1992*	*Programmed, 1992*	*Combat, 1992*
Strategic nuclear retaliation (day-to-day alert)	Damage expectancy against 5,274 targets	.6	.4	.5	.7
Theater nuclear retaliation, Europe (generated alert)	Damage expectancy against 1,513 targets	.6	.6	.5	.9
Simultaneous defense of:					
Central Europe (the M + 90 threat)	Probability of maintaining a forward defense	.2	.2	.3	.5
Persian Gulf (the M + 90 threat)	Probability of a forward defense of Saudi Arabia	.4	.3	.4	.6
Simultaneous deployments by M + 90 to:					
Central Europe	Thousands of short tons delivered	648.0	648.0	1,320.8	1,560.7
Persian Gulf	Thousands of short tons delivered	365.1	365.1	365.1	821.4
Simultaneous resupply by M + 90 of:					
Central Europe	Millions of short tons delivered	14.8	14.8	15.5	17.6
Persian Gulf	Millions of short tons delivered	1.9	1.9	2.8	2.8
Attack by three carrier battle groups on Murmansk	Probability of destroying Soviet Northern Fleet	[a]	[a]	.2	[a]

Sources: Author's estimates; and tables 5-9, 5-10, 5-11, 5-12, 5-13.
a. No carriers available for mission.

By contrast, the combat force equals or exceeds the performance of the programmed force except in two instances. Under the assumptions made about carrier operations, the combat force has no carriers available for an early attack on Murmansk or Vladivostok, and it delivers less tonnage to Europe prior to the very short-warning attacks by the Warsaw Pact at M + 4 and M + 9. Otherwise, however, it goes well beyond restoring the level of performance achieved by the baseline force in 1981. The combat force has a 50 percent chance of maintaining a forward defense in Central Europe against the worst threat (a large-scale attack at M + 90) and a 60 percent probability of blocking a Soviet attack north of Saudi Arabia, even assuming that the United States had not responded earlier to the occupation of northern Iran by the USSR. What is more, during the five-year planning period the combat force costs cumulatively $230 billion less (in outlays) than the programmed force.

The Reasons for the Differences

On the face of it, such a disparity in cost and effectiveness between the programmed force and the combat force may lack the ring of plausibility. However, a close look at the composition of the two forces and their performance in the standard contingencies makes clear why the differences exist and why they are so large. Perhaps the most startling conclusion is that the combat force's strategic offense will do 25 percent better than the programmed force's against the identical threat and target list (shown in table 5-9), yet cost cumulatively $80 billion less in budget authority. One reason for this disparity is that the Reagan administration, to stay within the MIRV limits set by the SALT II agreement, is assumed to continue retiring Poseidon submarines as new Trident boats enter service instead of deactivating the more vulnerable Minuteman III ICBMs. By contrast, the combat force planners keep the Poseidons and retire the Minuteman IIIs. If the administration were to worry more about second-strike deliverable warheads and less about retention of the Triad and the alleged difficulty of communicating with submarines at sea, it would change this policy, in which case the programmed force could have a somewhat higher damage expectancy than the combat force. However, even with this change, the large difference in cumulative costs would remain.

In part this difference exists because it is assumed that the adminis-

Table 5-9. Performance of the Three Forces in Strategic Nuclear Retaliation, Day-to-Day Alert, 1981, 1992

	Force			
	Baseline			
Targets and weapons	*1981*	*1992*	*Programmed, 1992*	*Combat, 1992*
Targets[a]	5,274	5,274	5,274	5,274
Weapons				
On launchers	9,592	9,592	12,516	11,058
Alert surviving weapons[b]	4,595	3,722	4,317	4,491
Delivered weapons	3,565	2,432	2,744	3,525
Targets destroyed	3,278	2,236	2,719	3,451
Damage expectancy (percent)	62	42	52	65

Source: Author's estimates.

a. Targets consist of: 1,598 hard strategic targets; 752 soft strategic targets; 640 peripheral attack forces (medium-range missiles and bombers) targets; 404 bases for ground, air, and naval general purpose forces; 500 noncollocated energy and transportation targets; and 1,380 urban-industrial aiming points in 200 cities.

b. After a well-executed Soviet surprise attack. Only surviving ICBMs, bombers on ground alert, and ballistic missile submarines on station are defined as having alert surviving weapons.

tration will not want to risk being criticized for reducing the number of strategic nuclear delivery vehicles without having received some compensation from the USSR. Therefore the administration will keep a number of B-52 and FB-111 bombers in the force even though, because of SALT II constraints, these bombers cannot be equipped with long-range air-launched cruise missiles (ALCMs). Such a decision means that, while the cost of the bomber force remains high, the number of alert bombers remains low. Furthermore, their ability to deliver short-range attack missiles and gravity bombs is even lower because they must penetrate Soviet defenses greatly improved by the deployment of good look-down, shoot-down capabilities. Still worse, the administration's expensive B-1Bs (canceled by the combat force planners) add little to the damage expectancy already achieved by SLBMs and ALCMs in a retaliatory strike. The cost-effectiveness of B-1Bs is so poor because their alert rate is low and their penetration probability is lower than that of the much cheaper cruise missiles. Precisely for this reason, the combat force planners replace the B-52s with relatively inexpensive cruise missile carriers, each with a payload of 28 ALCMs. Consequently, the bomber component of the combat force covers as many targets as the programmed bombers, but at much lower five-year costs.

Neither force has deployed the advanced technology bomber (Stealth) by 1992. In fact, the combat force planners stretch out its development and fund the program at lower levels than the administration, largely because, by 1992, the air-launched cruise missile is assumed to have good reliability (80 percent), an excellent probability of penetrating improved Soviet defenses, and high accuracy against both hard and soft targets.

The two forces deploy the same number of Trident submarines and Trident II (D-5) ballistic missiles, which are credited with a good hard-target-kill capability. However, the administration not only retains 550 Minuteman III and 450 Minuteman II ICBMs; it also deploys 100 MX missiles, as originally planned. However, only 4 percent of the ICBMs survive a Soviet first strike; consequently, they contribute little to the coverage of enemy targets compared with the SLBMs and ALCMs. The combat force planners also retain ICBMs: 416 Minuteman IIIs and 450 Minuteman IIs. This is done because their operating and support costs are low and because their retention complicates any Soviet attempt, in a surprise attack, to eliminate both the ICBMs and the alert bombers.

As is evident, the administration obtains a much lower return than the combat force planners do on investment in bombers and missiles, measured by the cost of each delivered warhead. As a matter of fact, the cost of destroying targets on a second strike with the B-1B and the MX is so high (because of their vulnerability) that these systems are of questionable worth even as bargaining chips. Indeed, it is an open question whether the Russians would trade much for the cancellation of these two systems or whether the resulting benefits in arms control would be worth the billions invested in the bargaining chips.

Another reason why the administration gets such a relatively low return on investment compared with the combat force planners is because the administration puts much larger funding into relatively unproductive active defenses against bombers and missiles. For example, the administration acquires F-15s and OTH-Bs (over-the-horizon backscatter radars) for the upgrading of U.S. antibomber defenses. Yet this investment in no way diminishes the damage inflicted by the Soviet first strike, which is executed with ICBMs, and it has no impact on any follow-on attacks because the Soviet ICBMs are also able to eliminate both the radars and the F-15 bases as part of the initial strike. Less susceptible to analysis is the administration's investment of more than $30 billion in the Strategic Defense Initiative (more familiarly known as Star Wars), simply because this initiative will produce nothing worth deploying by 1992. However, the combat force planners cut this total to $10 billion on the grounds that the development of a prototype ballistic missile defense facility at Kwajalein Island in the Pacific and the determination of which of the more exotic technologies may be worth developing should cost about a third of what the administration is proposing. None of these steps would risk violating the Anti-Ballistic Missile Treaty or require its modification.

Even though the tactical nuclear capability deployed by the combat force planners is much smaller, measured in warheads, than the capabilities in the programmed and the baseline forces, it provides a much greater damage expectancy and costs somewhat less than the programmed capability (table 5-10). However, this difference is by no means a necessary one. The combat force planners phase out all the nuclear weapons designated for delivery by artillery tubes and strike aircraft, but they make up for this reduction in two ways. First, they use all the Pershing II ballistic missiles, ground-launched cruise missiles, and Tomahawk nuclear land-attack missiles based on attack submarines against tactical targets; consequently about the same number of warheads survive a Soviet attack as in the other two forces, which rely heavily on vulnerable aircraft and short-range delivery systems for the coverage of battlefield and interdiction targets. Second, the combat force planners allocate a fairly small number of relatively high-yield weapons to barrage attacks on the forward echelons of the Warsaw Pact's ground forces. Because of these two measures, the combat force planners get a factor of four greater target coverage than the administration. With similar changes, which no doubt would produce strenuous resistance from the Army and the Air Force, the programmed force could obviously produce identical results.

The combat force does not demonstrate such striking increases in effectiveness with its conventional capabilities, but they are more significant. With allied forces, these capabilities provide the noncommunist world with a reasonably sturdy deterrent to conventional attacks by the USSR and its satellites. Yet, cumulatively over the five years, they cost $200 billion less in budget authority than the programmed force's nonnuclear arsenal.

The reasons for these differences are not hard to find. The programmed forçe invests heavily in expensive aircraft for long-range air superiority and interdiction, expensive ships for carrier battle groups and amphibious warfare, and expensive widebodied aircraft for intercontinental mobility. Yet the returns on these investments are uniformly poor.

In the Central Region, the forward stockage of supplies by the Warsaw

Table 5-10. Performance of the Three Forces in Tactical Nuclear Retaliation in Central Europe, Generated Alert, 1981, 1992[a]

	Force			
	Baseline			
Targets and weapons	*1981*	*1992*	*Programmed, 1992*	*Combat, 1992*
Targets[b]	1,513	1,513	1,513	1,513
Weapons				
On launchers	7,000	7,000	5,655	915
Alert surviving	1,605	1,605	1,334	775
Delivered	1,173	1,173	963	620
Targets destroyed	939	939	770	1,407
Damage expectancy (percent)	62	62	51	93

Source: Author's estimates.

a. It is assumed that the USSR launches its nuclear attack during a conventional war and that all nuclear launchers are on a high (or generated) alert.

b. Targets consists of: 72 main air bases in Eastern Europe; 191 transportation choke points; 162 command and storage bunkers; 1,088 enemy battalions along a front of 750 kilometers.

Pact combined with the density of the rail and road network (as well as petroleum pipelines) in Eastern Europe makes it difficult to bring about any meaningful reduction in the tonnage delivered to the fighting forces of the Pact in fewer than two weeks. This is true even with the increasingly sophisticated aircraft and smarter munitions being acquired by the U.S. Air Force for deep air superiority and interdiction. Yet this is the period in which the most critical land battles are likely to be fought. In these circumstances, F-111s, F-15Es, and F-16s could be used for close air support. But a much greater payoff would be obtained (shown in table 5-11) by reducing the acquisition of F-16s and F-15Es, installing a barrier along the inter-German border (a proposal resisted by the Federal Republic of Germany), and buying more close air support aircraft for use in conjunction with allied ground forces to attack the enemy's forward concentrations, however distasteful such a trade might be to the Air Force.

The three new carrier battle groups and the added brigade's worth of amphibious lift for the Marine Corps appear to be excessive for the contingencies in which these forces might reasonably be involved. As for attacks on Murmansk or Vladivostok, the amphibious forces would probably have no role to play at all, and three carrier battle groups are not nearly sufficient to achieve a significant probability of crippling Soviet naval forces in port (table 5-12). The U.S. Marines, admittedly, now have the mission of operating in North Norway. But this does not

Table 5-11. Performance of the Warsaw Pact and the Three Forces in the Defense of Central Europe, 1981, 1992

	Warsaw Pact attacks at:			
Force	*M + 4*	*M + 9*	*M + 14*	*M + 90*
Warsaw Pact				
Divisions	30	57	90	110
Close air support aircraft	300	300	300	300
Baseline (1981)				
Divisions	30⅓	43	46	49
Close air support aircraft	600	600	600	600
Probability of maintaining a forward defense	.63	.46	.29	.21
Baseline (1992)				
Divisions	30⅓	43	46	49
Close air support aircraft	600	600	600	600
Probability of maintaining a forward defense	.57	.39	.24	.17
Programmed (1992)				
Divisions	30⅓	43	46	49
Close air support aircraft	600	600	600	600
Probability of maintaining a forward defense	.73	.58	.40	.32
Combat (1992)[a]				
Divisions	30⅓	43	46	62⅔
Close air support aircraft	600	1,252	1,252	1,252
Probability of maintaining a forward defense	.73	.72	.52	.54

Sources: Author's estimates; and tables 5-2, 5-3, 5-14.

a. The combat force is not as sophisticated as the programmed force. However, it benefits from a barrier along the interzonal frontier, more divisions by M + 90 (from the U.S. National Guard and Reserve forces), and a doubling of close air support aircraft.

mean either that they have to be delivered there by amphibious assault, or that their arrival needs to be supported by carrier-based aircraft. A more productive investment than this expensive emphasis on fragile "power projection" capabilities would be in additional escorts for convoy duty. And to the extent that it is appropriate to engage in offensive operations in the Barents Sea or the Sea of Japan, attack submarines are the most efficient instrument for that purpose.

The Air Force plan to double airlift capacity from 32.4 million ton-miles a day to 66.0 million ton-miles sounds impressive. But this increase makes little difference in the outcome of short-warning attacks in Europe (at M + 4 and M + 9), especially when a barrier is installed and Warsaw Pact forces are properly evaluated. The increase barely satisfies the deployment "requirement" for the M + 14 case; and it is not sufficient to move the tonnages involved in major and simultaneous deployments to Europe and the Persian Gulf. Should two such contingencies arise, only the eight SL-7 fast sealift ships acquired by the Carter administration

Table 5-12. A U.S. Attack with Carrier Battle Groups on the Murmansk Area

	Beginning of:					
U.S. force and performance	*D-day*	*D+1*	*D+2*	*D+3*	*D+4*	*D+5*
Three carrier battle groups						
Probability of force surviving	.24	.10	...	...	...	...
Percent of U.S. ships surviving	100	32	...	...	...	...
Percent of Soviet ships sunk	...	7	9	...	...	...
Six carrier battle groups						
Probability of force surviving	.39	.32	.23	.08	...	...
Percent of U.S. ships surviving	100	66	37	11	...	...
Percent of Soviet ships sunk	...	13	22	27	29	...
Nine carrier battle groups						
Probability of force surviving	.49	.48	.47	.45	.42	.38
Percent of U.S. ships surviving	100	77	59	45	33	23
Percent of Soviet ships sunk	...	20	36	48	57	63
Ten carrier battle groups						
Probability of force surviving	.51	.52	.53	.54	.57	.60
Percent of U.S. ships surviving	100	80	64	52	43	36
Percent of Soviet ships sunk	...	22	40	54	66	75

Sources: Soviet force size and composition based on Department of Defense, *Soviet Military Power*, pp. 8, 13, 107; other data based on author's estimates.

would be available for deployments to Southwest Asia, assuming all the airlift is used to reinforce Europe (table 5-13).

This dependence on expensive airlift arises to a degree from the exaggerated threats postulated for the three short-warning contingencies in Europe. But it is also a function of the entrenched military belief that U.S. and allied civilian leaders cannot be trusted to make prompt deployment decisions based on early warning that may be precise about Soviet mobilization and deployment moves but ambiguous about the intentions of the Kremlin. This belief may have some merit to it, even though the recent evidence for it is mixed. But rather than continue with the pretense that Soviet forces are capable of mobilization and deployment feats that no one else knows how to perform, it would be preferable to explain to U.S. and allied leaders that airlift is not their only option. Major obstacles along the inter-German border, for example, would make a surprise attack more difficult and necessitate bigger and longer buildups by the Warsaw Pact to overcome them. Moreover, barriers are cheap.

By the same token, fast sealift costs much less per ton-mile than airlift and, with a week of preparation, can deliver large tonnages to the major northern ports of Europe in another week. Ships such as the SL-7 can

Table 5-13. Simultaneous Deployments of the Three Forces to Central Europe and the Persian Gulf, 1981, 1992

Thousands of short tons delivered

Force	*M + 4*	*M + 9*	*M + 14*	*M + 45*	*M + 90*
Baseline (1981, 1992)					
Central Europe	28.8	64.8	100.8	324.0	648.0
Persian Gulf	...	...	...	121.7	365.1
Programmed (1992)					
Central Europe	58.7	132.1	205.5	660.4	1,320.8
Persian Gulf	...	...	...	121.7	365.1
Combat (1992)					
Central Europe	28.8	64.8	678.9	1,236.7	1,560.7
Persian Gulf	...	...	...	273.8	821.4

Sources: Table 5-11; and author's estimates.

reach the Persian Gulf in two weeks. Airlift, in other words, is not the only solution to rapid deployments or to the very short warning cases that receive such emphasis in current planning. It is not competitive at all with sealift when it comes to the longer, larger buildups that would accompany or follow a major international crisis and that are, in reality, the most probable and most threatening eventualities.

The combat force exploits all these possibilities. It also acquires the capability to deal with the M + 90 case in Europe and to respond simultaneously in the Persian Gulf as well as reinforce U.S. units in South Korea. The force does so by adding to the large but not very productive investment that already goes into the National Guard and Reserve forces. As of now, some of these forces could and probably would deploy along with active-duty ground and tactical air units. But a large percentage of the Army Guard and Reserve is unlikely to be ready to move overseas in fewer than ninety days, because of their need for reorganization and large-unit training. Moreover, most of these forces, whether Army or Air Force, have largely hand-me-down equipment and shortages of that. The combat force furnishes them with additional modern weapons, with particular emphasis on close air support and fighter aircraft for Air Force squadrons. The force also provides for the deployability of all combat units by M + 90 through increased amounts of large-unit training.

The additional ground forces, the major increase in Air Force squadrons equipped and trained with close air support aircraft, and the introduction of a major barrier along the inter-German frontier are the crucial ingredients in enabling the combat force to outperform the programmed force in all the nonnuclear planning contingencies, even though the combat force designers reject many of the investments made by the administration. More specifically, the combat force general purpose navy is smaller than the programmed fleet by some sixty ships, mostly because it does not buy the three new carrier battle groups, the additional amphibious lift for a Marine brigade, or the surface action groups that center on rehabilitated battleships. However, the combat force's twelve carrier battle groups and amphibious lift for three Marine amphibious brigades are sufficient to deal with the planning contingencies requiring maritime power projection. Furthermore, the combat force navy contains enough destroyers and frigates to protect nine large convoys a month, the number needed to support the U.S. ground and tactical air forces deployed to Europe, the Persian Gulf, and Korea.

Combat force planners depend on the units already deployed in peacetime, prepositioned matériel, existing airlift and the civil reserve airlift force, and barriers in place to deal with the surprise attacks that could occur in Europe at M + 4 and M + 9. For other contingencies in Europe and for simultaneous deployments to the Persian Gulf and Korea, they acquire a large fleet of fast sealift ships with a roll-on-roll-off capability that permits the discharge of cargoes over the beach. The combination of this expanded intercontinental lift with the improved National Guard and Reserve forces means that the combat force is able to deploy a much larger capability overseas than the programmed force during the first several months of a crisis (as shown in table 5-14), even though the active-duty component of its conventional posture is smaller and costs substantially less.

Not only does the active-duty combat force have fewer divisions, tactical fighter squadrons, and battle forces ships; it depends somewhat less for its performance on the most sophisticated equipment, which is a major factor in the cost of the programmed force. Consequently, the combat force needs fewer active-duty personnel, incurs lower costs for operation and maintenance, and can manage with a more modest budget for military construction (including housing for military families). Its allocations for research, development, and procurement are also much lower not only because they shun the high-cost, low-yield capabilities and slow the pace of modernization, but also because they reduce the rate of acquisition of expensive modern munitions.

Despite the importance of sustainability, there are several grounds

Table 5-14. M + 90 Deployment of the Programmed and Combat Conventional Forces, Fiscal Year 1992[a]

Planning contingencies and forces	Divisions	Marine amphibious forces[b]	Tactical fighter squadrons: Air superiority and interdiction	Tactical fighter squadrons: Close air support	Anti-submarine warfare barriers[c]	Escorted convoys per month	Carrier battle groups	Airlift aircraft	Fast sealift ships
Central Europe									
Programmed	15	. . .	48	18	. . .	. . .	. . .	524	. . .
Combat	23	. . .	43	45	. . .	. . .	. . .	304	20
North Norway									
Programmed	⅔	⅓	. . .	8	. . .	. . .	3	. . .	. . .
Combat	⅔	⅓	. . .	8	. . .	. . .	3	. . .	. . .
Mediterranean									
Programmed	. . .	. . .	. . .	. . .	. . .	. . .	2	. . .	. . .
Combat	. . .	. . .	. . .	. . .	. . .	. . .	. . .	. . .	. . .
Atlantic and Caribbean									
Programmed	. . .	⅓	. . .	. . .	1	5	. . .	. . .	. . .
Combat	. . .	. . .	. . .	. . .	1	5	. . .	. . .	. . .
Persian Gulf									
Programmed	3⅔	⅓	6⅔	8⅓	. . .	. . .	3	. . .	8
Combat	8⅔	⅓	24⅔	8⅓	. . .	. . .	3	. . .	18

Indian Ocean									
Programmed	...	...	...	...	1	1	...	...	...
Combat	...	...	...	...	1	2	...	...	...
Korea									
Programmed	1⅔	...	12	...	...	...	...	...	...
Combat	2⅔	...	17	...	...	...	...	...	...
Pacific									
Programmed	...	½	...	...	1	2	3	...	...
Combat	...	⅓	...	...	1	2	3	...	...
Alaska									
Programmed	⅓	...	3	...	...	...	...	...	...
Combat	⅓	...	3	...	...	...	...	...	...
Panama									
Programmed	⅓	...	3	...	...	...	...	...	...
Combat	⅓	...	3	...	...	...	...	...	...
CONUS[d]									
Programmed	15⅓	...	48	...	...	...	4	...	...
Combat	1⅓	...	3	...	...	...	3	...	...
Total									
Programmed	37	1½	120⅔	34⅓	3	8	15	524	8
Combat	37	1	93⅔	61⅓	3	9	12	304	38

Sources: Table 5-6; and author's estimates.
a. All planning contingencies are assumed to occur simultaneously.
b. A marine amphibious force comprises the ships, personnel, equipment, and supplies for one Marine Corps augmented division.
c. These consist of mines, attack submarines, and land-based patrol aircraft.
d. Continental United States. Forces that cannot be deployed by M + 90 or that are in training, maintenance, or overhaul.

for imposing such a slowdown. First, the technology of modern munitions is undergoing rapid change, which means that existing war reserve stocks will incur quick obsolescence. Second, major uncertainties exist about the extent to which the USSR can sustain one or more conflicts of high intensity, but with the evidence pointing toward a Soviet capability for short rather than lengthy campaigns. Third, even greater uncertainties exist about just how large the U.S. war reserve stocks of smart munitions should be. To a considerable degree, the services are inclined to acquire modern munitions pretty much on the same basis they use to buy the more traditional dumb munitions, of which they have large stocks. That is to say, they compute the average daily firing rates of their various weapons and establish a stock of "required" munitions measured in the number of days sufficient to keep their units supplied until they can be sustained from the output of expanded production lines. The objection to this approach is that if the munitions are as smart as they are supposed to be, calculating stockpiles in days of supply rather than in targets and kill probabilities is extravagant; whereas if they are not all that smart but continue to be expensive, the dumb but much cheaper munitions remain the better buy. These issues must be resolved before the Defense Department spends well over $100 billion on large stocks of modern munitions, which will then have to be replaced by a newer and more effective generation of missiles and shells. Hence the more cautious policy adopted by the combat force planners and the lower funding of modern munitions.

CHAPTER SIX

WHAT HAS GONE WRONG

IF THE COMBAT force is so superior to the programmed force in performance, but costs so much less, why in the world have not the professionals in the Pentagon bought it rather than the programmed force? A possible answer is that the superiority of the combat force is a function of biases in the analysis rather than of its greater merit. And it is certainly true that the calculations used to determine the outcomes of hypothetical conventional conflicts make it very difficult and costly to compensate for increases in enemy numbers with increases in U.S. quality. But these are precisely the kinds of calculations used by Soviet as well as U.S. planners, and there is a great deal of experience to support them. In all respects, moreover, the three forces were tested under identical conditions. Consequently, the question remains: if not the combat force or something like it, why not?

The Role of the Secretary of Defense

A review of some history is necessary to arrive at the answer. When the Defense Department was created in 1947, it consisted of three services, including the newly created Air Force (or four if the Marine Corps is counted as separate from the Navy), the Joint Chiefs of Staff—a kind of cabinet in which the chairman has become slightly more powerful than his colleagues but much less powerful than a British prime minister—and a secretary of defense as its titular head, but without the authority or knowledge to settle the running disputes among the services and within the Joint Chiefs of Staff. Despite frequent proposals for unifying the services and putting them into one uniform, it was always evident that the resistance to such a revolution, particularly from the Navy, would be bitter and costly. It was equally clear that even with unification the same issues about the conditions of deterrence, the kinds of wars to prepare for, roles and missions, and the allocation of resources would continue among the proponents of nuclear and conventional capabilities, ground and tactical air forces, land-based and sea-based units, infantry and armor, bombers and fighters, and carriers and submarines. In recognition of these hard realities, President Dwight D. Eisenhower chose to give increasing power to the secretary of defense, so that by the end of his second term it was the secretary of defense who could not only set budgets and settle disputes but also take initiatives in all those areas of force size and composition, missions, and resource allocation that the Joint Chiefs of Staff had come to regard as their exclusive preserve.

Robert S. McNamara, secretary of defense from 1961 until 1968, was the first to exercise this power to its full measure. He was able to do so mostly because of his own energy, courage, and analytical abilities, the force of his personality, the knack of being able more often than not to dictate the terms of a debate, the dedication of his staff and, most important of all, the support of the president. Without that support it is difficult for any secretary of defense to accomplish a great deal in the face of opposition from the services, the Joint Chiefs of Staff, and influential members of Congress.

Even with that support the battles over plans, programs, and budgets can be long and hard. They proved to be especially intense during the McNamara years. The military unsuccessfully resisted what it regarded as the intrusion of the secretary and his staff into force planning and programming. For the most part the military succeeded in fending off what it saw as civilian meddling in military operations, although it was forced to accept constraints on its activities during the missile crisis and in Vietnam, as it had in Korea. The military also had to suffer the indignity of presidential participation in the details of bombing raids against North Vietnam. Congress, initially bemused by McNamara's encyclopedic knowledge and awed by his grasp of the issues, increasingly resisted what it considered his rough treatment of its allies in the services, his lack of sympathy for special interests, his unwillingness to cut deals with individual members, and his resistance to traditional divide-and-rule tactics.

Despite this opposition, McNamara undoubtedly bequeathed his successor a more rational, more powerful, and more efficient military capability than he had inherited in 1961. He also left a set of institutions and methods by which future secretaries could exert their authority in an orderly and systematic way. To his credit, he was not content to make his own personal mark on the Defense Department; he also sought throughout his seven years to embed his decisionmaking reforms into the structure of the department.

Planning, Programming, Budgeting

At the heart of these reforms was the planning-programming-budgeting system (PPBS). This system is both a process and, what participants in it frequently forget, a method. Each can endure without the other, but it was McNamara's intent that they should be combined so as to link resources with objectives and the forces to reach them. The PPBS, in effect, was meant to persuade policymakers to think simultaneously about the goals they wanted to accomplish and how much it would cost to reach their goals.

As designed, the system accomplished other purposes as well. Essentially, what triggered its operation was a series of papers on major force planning issues, which came to be known as draft presidential memoranda (DPMs). These memoranda originated in the office of the secretary of defense and were scrutinized by the secretary before being distributed for comment from interested military and civilian parties. That ensured two consequences: first, that the secretary held the initiative in proposing changes in plans and programs; and second, that the issues were approached systematically and analyzed, as much as possible, according to the criteria of cost and effectiveness.

Before preparation of the memoranda, the secretary's staff (which included military as well as civilian planners) exchanged information and views with the services and the organization of the Joint Chiefs of Staff, and frequently based their analyses on studies done by the services at the request of the secretary. But despite a good deal of collaboration, the DPMs remained the secretary's memoranda. They forced the rest of the department to respond to and comment on his proposals. Furthermore, if the commentators were to obtain a hearing, they would have to discuss the issues in the analytical language of the memoranda. It was hoped that this insistence would raise the level of the debate in the department and induce the services to be explicit about objectives and to carry out their own cost-effectiveness studies.

McNamara followed his predecessors in pursuing some piecemeal unification with the establishment of such agencies as the Defense Intelligence Agency, the Defense Logistics Agency, and the Defense Nuclear Agency, all of which were intended to consolidate functions duplicated in each of the services. But he relied mostly on the PPBS to resolve interservice and intraservice disputes, avoid duplication, control the traditional tendencies of the three services to work at cross-purposes, and curb the propensity of each to prepare for its preferred kind of war. To the same ends, he introduced the program budget that showed the resources of the department broken down by major missions and functions. Thus, all the strategic nuclear forces and activities directly associated with them were placed in one program regardless of which service owned them. What was even more important, each DPM looked at, compared, and recommended programs according to the mission to be performed and the objective to be achieved. If ballistic missiles were more efficient than bombers, and SLBMs a better bargain than ICBMs, these considerations rather than a service's preference would dominate the allocation of resources.

The five-year defense program (FYDP), with its projection of forces and costs, was also introduced as an aid to the long-term planning essential to rationality in defense. McNamara liked to say, in this regard, that while he proposed to buy only what was needed, the need would determine the budget rather than the other way around. He found it useful at least initially to have analyses done that went beyond the standard cost-effectiveness comparisons to proposals for programs that were not constrained by a particular annual budget. These analyses expanded his range of choice and provided him with the room to bargain with the military and members of Congress. Once budget constraints were imposed, such analyses let him see what he could give up without damaging his essential purposes. Eventually, however, the step from plans and programs to specific budgets and allocations of resources would have to be taken, and that step could never constitute a simple adding up of all the programs recommended in the first drafts of the draft presidential memoranda. Fiscal constraints undoubtedly existed then as before and since, and choices had to be made. Indeed, from fiscal 1961 until fiscal 1966 (when the war in Vietnam began to have a serious impact on the defense budget), the average annual real increase in budget authority was less than 5 percent, despite the fiction that programs dictated budgets rather than the other way around.

Centralized Planning

The military has often been described as a state within a state, and there is more than a grain of truth in the description even in the United States. The Defense Department has its own territory spread in patches around America and the rest of the world; it has a population of more than 2 million people in uniform, and it exercises some control over another million civilian employees. It has its own laws and judicial system, and it has its own chief executive in the person of the secretary of defense. Within this sprawling country, planning, programming, and budgeting can be done with the utmost elegance, but if the plans and programs cannot be implemented as intended, all that elegance will have gone for naught.

The defense state is understandably controlled to a large extent by its military bureaucracy, and within that hierarchy not only is civilian control of the military a fundamental principle (as it is a constitutional requirement); subordinates within the chain of command are also trained to obey orders. At the same time, the military is made up of many constituencies, each convinced of the importance and rightness of what it is doing, each making a claim on the resources of the department, and each wanting to dispose of these resources as it sees fit. Accordingly, resistance to centralized planning is inevitable and so is a temptation to execute unpopular decisions in a perfunctory way. Indeed, some days a secretary of defense may wonder whether he controls more than one of the five rings of offices in the Pentagon.

Inevitably, these frictions—which cannot be wished or ordered away—raise questions about how best to plan and how best to ensure execution of the plans. The case for centralized planning remains persuasive, despite the criticism of it and the resistance to its continuation. The days have long since ended when the Army and Navy could each go its own way, and the only issues were how much money they should get and how it should be divided. Technology, if nothing else, ensured that result. Prior to World War I, armies owned the land and navies owned the seas, although each had made modest incursions into the domain of the other. By the end of World War I, the airplane had demonstrated that, whatever the platform from which it was launched, it could substitute for ground forces in destroying opposing ground forces and for battleships in destroying opposing battleships. By the end of World War II, the evolution of technology and the need to engage in coordinated, combined operations in every theater of war had forced the institution of unified commands to centralize the planning and control of ground, naval, and air forces. With the subsequent appearance of nuclear weapons, the development of cruise and ballistic missiles, the creation of the Air Force as a third service, and the prospect of space-based capabilities, it became increasingly evident—first of all to the services themselves—not only that new missions and new ways of performing old missions had broken down the old definitions of service responsibility, but also that traditional ways of thinking about force planning might have to be changed. The Army might argue that it should control everything based on land, but it had already conceded the independence of land-based air forces. The Navy was to be more successful in claiming that it should own anything based at sea, but it could see that the long-range bombers of the Air Force might one day challenge its monopoly of naval targets and that large, widebodied aircraft could reduce the need for both sealift and command of the seas. The Air Force was able to gain control of the ICBMs, but it could not find the logic to preserve a monopoly of nuclear weapons and its claim to the exclusive use of space was lacking in any kind of provenance. It is hardly surprising, then, that the Joint Chiefs of Staff became the arena for battles rather than the forum for decisions, and that President Eisenhower, who was a veteran of these disputes, chose to vest so much power in the secretary of defense. Only someone other than the interested parties could evaluate competing claims with some objectivity, assign (and reassign) missions, decide on risks, and allocate resources, and that someone had to be the civilian head of the department. Those functions, in fact, are what civilian control is all about in the United States today. The man on horseback is not the modern threat; the danger comes from anarchy and inefficiency.

Despite the imperatives of centralized planning, the secretary of defense can choose to exercise his authority in different ways. Like McNamara, the secretary can attempt to impose ideas on the military hierarchy in the almost certain knowledge that resistance will ensue. Like one successor, James R. Schlesinger, the secretary of defense can try to devise programs that will give the services a strong incentive to adopt and implement them, but with the risk that he will have to sacrifice efficiency to gain compliance. But whatever the tactics, the basic choice is between central planning and direction, on the one hand, and the free enterprise of the services, valuable in a market economy but out of place in the Defense Department.

The Decline of the Planning-Programming-Budgeting System

When Melvin R. Laird became secretary of defense in 1969, he seemed to believe otherwise. Although he did not abandon the PPBS, he responded to the military criticisms of McNamara by instituting what he called participatory management and by giving the initiative to the services in the planning cycle. The DPMs disappeared. The program objective memoranda (POMs) took their place. The services were their authors and the office of the secretary of defense now wrote issue papers to comment on them. After a little more than a year, however, Laird discovered that participatory management and the POMs meant that the priorities of the individual services were replacing the preferences of the president. To regain control over the process, he not only established budgetary constraints at the outset of the annual planning cycle, but also had his staff in the office of the secretary of defense produce a document known as the defense guidance, which provided a framework of objectives and forces within which the services would develop their POMs.

This was a looser rein than McNamara had used, and subsequent secretaries of defense found it a convenient way to exercise control while giving the services more of a role in the budgetary process. But the defense guidance, like the DPMs, could still be specific enough to ensure that individual service programs complemented rather than duplicated one another, and that the services emphasized national rather than their own purposes.

As is well known, the United States is a government of laws, not of men and women. Still, men and women write the laws and implement them. The PPBS can provide the secretary of defense with the tools for imposing coherence, discipline, and efficiency on the programs of the Defense Department. But the system can also serve as a facade behind which each service goes its own way. The facade is easy to maintain; application to the issues, long hours, and difficult choices are needed if the force planning system is to work as intended. The current secretary of defense apparently finds the PPBS less useful than the advice of the three services. The PPBS remains in place as a process (summarized in table 6-1), and an analytical office lives on in the office of the secretary of defense; more of the principals and more of the staff are engaged in the process than ever before. The defense guidance makes its annual appearance; the services issue their POMs. The office of the secretary produces its issue papers in response to the POMs; the five-year defense program purports to chart the future. As a further refinement, the capabilities being acquired during the FYDP are compared with what is presumably considered to be the ideal force. And what is that ideal force? It is the Joint Chiefs of Staff minimum-risk force (shown in table 6-2), a well-known compilation of what each of the services thinks would be desirable. Thus the Army dreams of twenty-five active divisions, the Navy of twenty-two carrier battle groups, and the Air Force of thirty-eight active tactical fighter wings. The Marine Corps alone does not have to dream; the law stipulates that now and forever it must have a minimum of three divisions. In other words, despite all the machinery, despite all the bloated committees, it is the objectives of the individual services that still dominate force planning.

Inefficiencies

Admittedly, the defense budgets provided by the Reagan administration, though generous to a fault, have not proved sufficient to fund the minimum-risk force in the foreseeable future. But the budgets were sufficiently generous for the services to be able, on the installment plan, to begin the fulfillment of most of their wishes. The secretary and his deputy denied the services little; at worst, they limited the number of weapons and equipment the services could buy and the speed with which they could develop new systems to replace them. The upshot of this tolerance is that all three services are trying simultaneously to expand their capabilities, upgrade older weapons, and replace them as rapidly as possible with new and more costly models. Furthermore, in doing so, each is investing in weapons that will enable it to operate independently of the others. The Army is buying expensive attack helicopters and air defense weapons because it does not expect to be given the necessary support by the Air Force. The Air Force, which could acquire more close air support aircraft and short-range air defense interceptors, prefers to invest in long-range fighter-attack aircraft that can attack targets deep in the enemy's rear and conduct an interdiction

Table 6-1. Summary of the Current Planning-Programming-Budgeting System

Month	*Year*	*Document*	*Purpose*
October	1	Joint strategic planning document (JSPD)	Summary of Joint Chiefs' force levels required to execute national strategy with "reasonable assurance" of success
January	2	Defense guidance (DG)	Guidance by the secretary of defense on policy, strategy, force planning, resource planning, and tentative fiscal constraints
May	2	Program objective memoranda (POMs)	Programs for investment, operations, and support proposed by the Army, Navy, and Air Force
June	2	Joint program assessment memorandum (JPAM)	Assessment by the Joint Chiefs of the capabilities and risks inherent in the composite POM forces
August	2	Program decision memoranda (PDMs)	Programs approved by the secretary of defense as the basis for the defense budget
December	2	Draft defense budget	Defense budget proposed to the president for submission to Congress (after review by the Office of Management and Budget)
January	3	Federal budget	President's request to Congress for defense and other appropriations

Source: Joint Department of Defense–General Accounting Office Working Group, *The Department of Defense's Planning, Programming, and Budgeting System* (GAO, 1983), pp. 14, 35.

Table 6-2. The Joint Chiefs of Staff's View of Minimum-Risk Forces, Fiscal Year 1991

Force	*Active duty*	*Reserve*
Intercontinental ballistic missiles	1,254	. . .
Fleet ballistic missile submarines	44	. . .
Strategic bombers	483	. . .
Army divisions	25	8
Marine amphibious forces	4	1
Air Force tactical fighter wings	38	19
Aircraft carrier battle groups	22	. . .
Intercontinental lift aircraft	632	. . .
Intratheater lift aircraft	458	302

Source: *Armed Forces Journal International*, vol. 119 (August 1982), p. 38.

campaign in the hope of winning the war regardless of what happens to the Army. The Navy, asserting its independence of everyone else, prepares to fight its own small wars with amphibious forces and carrier-based tactical aircraft, more than half the cost of which goes into protecting this power-projection capability (shown in table 6-3).

Table 6-3. Average Cost of a Carrier Battle Group
Costs in billions of 1986 dollars

Component	*Number of ships or aircraft wings*	*Acquisition cost*			*Annual operating cost*		
		Total	*Offense*	*Defense*	*Total*	*Offense*	*Defense*
Attack carrier	12	38.0	38.0	...	5.7	5.7	...
Aircraft wings	12	57.5	38.3	19.2	14.4	9.6	4.8
Cruisers, destroyers	96	88.1	...	88.1	14.7	...	14.7
Mobile logistic ships	48	10.3	10.3	...	1.7	1.7	...
Frigates	48	10.3	...	10.3	1.7	...	1.7
Attack submarines	24	18.0	...	18.0	3.0	...	3.0
Auxiliaries	12	1.3	1.3	...	0.4	0.4	...
Total	252	223.5	87.9	135.6	41.6	17.4	24.2
Average battlegroup	21	18.6	7.3	11.3	3.5	1.5	2.0
Percent of costs for offense and defense	...	100	39	61	100	43	57

Source: Author's estimates.

The inefficiency of the nuclear force planning process is especially striking. The Army deploys the Pershing II, the Air Force the ground-launched cruise missile, and the Navy the nuclear Tomahawk—all intended for the same targets. The Air Force is deploying one new heavy bomber (the B-1B) and developing another (the advanced technology bomber) while it continues to make improvements in its older B-52s. At the same time, as it struggles to modernize its increasingly vulnerable ICBM force with the MX, and then Midgetman, the Navy prepares to acquire an excellent hard-target-kill capability with the Trident II (D-5) missile.

Exaggerated Threats

This kind of independence is costly. It is made more costly by the picture painted of Soviet military power, especially by the Defense Intelligence Agency, an organization that reports to the Joint Chiefs of Staff and engages in what is known as counterpart intelligence—a mode of estimating that requires each service to evaluate its counterpart in the USSR. As might be expected the estimates are staggering. Not only does the Soviet Union outnumber the United States in men, divisions, aircraft, missiles, ships, and submarines; it is rapidly catching up in the quality of its weapons, and in some technologies is said to have the lead. The future, therefore, is bound to be bleak unless the United States continues and increases its efforts. Soviet leaders may have been cautious in the past. But with so much military power already at their disposal, and with the end of U.S. nuclear superiority (always asserted but never defined), such restraint cannot be counted on in the future.

This pessimistic diagnosis inevitably leads to a demanding prescription. Since the United States and its allies supposedly cannot compete in numbers with the Red hordes (even though the United States alone probably managed to put more men under arms than the USSR in World War II), the only salvation is a technology so sophisticated that it permits U.S. weapons to overcome the alleged Soviet advantage in numbers with correspondingly high kill ratios. Thus if the Russians have five times more tanks, each new U.S. M-1 tank must be able to destroy at least five Soviet tanks for every M-1 that is lost. Because of the rapidly growing power of the Soviet forces, new U.S. equipment must be developed and deployed with increasing speed. Rather than fly before buy, the policy has become buy before fly. Concurrent development and acquisition must be the rule of the day, even if it results in costly mistakes, weapons with major defects, and repairs that must be made after the system has left the production line. As for the view of a decade ago that a "hi-lo" mix of weapons is the way to combine numbers with quality, that too must be discarded in favor of high technology all through the force, regardless of what it costs.

The Congressional Role

Such an attitude, combined with the independence of each service and a policy of laissez-faire at the top of the Defense Department makes adoption of something resembling the combat force a virtual impossibility. Hypothetically, Congress could force the department to make more rational choices. Realistically, it is most unlikely to do so. It is a consensual rather than an executive body; consequently, it can hardly be expected to enforce the kind of discipline that the executive branch has abandoned. There are, in addition, other inhibitions on congressional performance of that role. Defense now proposes each year to fund more than $100 billion of contracts that will be let or renewed in specific parts of the country. At a minimum, no member of Congress can ignore the economic consequences of these decisions; at a maximum, the temptation to encourage that funding, to regard it as a prudent alternative to the vagaries of private enterprise, and to claim a share of it for a particular state or district is powerful indeed.

It may be stretching a point to suggest that defense contracts have become the Works Projects Administration of the 1980s. But it is noteworthy that, despite the current pressure for a leaner defense, Congress remains reluctant to forgo or cancel major weapons systems on its own initiative. Its members even boast that they can cut the defense budget without affecting major procurement. And, of course, the Defense Department encourages congressional timidity. What will happen as a result of this abstinence, to essential ground support equipment, spare parts, ammunition, and fuel, is left unsaid.

Congress is hardly constituted to make these difficult decisions. Whatever its members may think of senior military and civilian officials in the Defense Department, they recognize that these officials represent a powerful constituency of active and reserve military personnel, organizations of retired military, contractors, and even labor unions which, if they agree on nothing else, agree that defense always needs more money. Such a constituency is not lightly defied or offended. There is the further risk that the Defense Department, as the keeper of the secrets and the guardian of military experience and judgment, will suddenly attack an errant member of Congress with previously undisclosed data and a parade of senior officers who will solemnly assert the absolute requirement for whatever has come under scrutiny.

The danger of public and devastating exposure is inhibiting enough. But the nature of the congressional process reinforces the resulting caution. Committee hearings are at the heart of the process, and they are more investigative and judicial than analytical in character. They do not lend themselves well to the complicated business of tracing the relationships among objectives, forces, and costs, but they are excellent vehicles for probing the safer and more exciting subjects of waste, fraud, and abuse. Members of Congress may not feel comfortable debating such issues as threats, contingencies, targets, and probabilities of success; these issues do not yield readily to lawyerlike questions and brief answers. Moreover, dealing with them calls for more than experience in the courtroom. A senator or representative, on the other hand, can feel at home probing charges of inflated overhead costs, excessively priced spare parts, and violations of contracts. These charges call for skills and knowledge over which the military cannot claim a monopoly or even expertise.

It is true, of course, that congressional staffs have multiplied during the last fifteen years. But while they may have become too large for the traditional investigations of Congress—and that is debatable—they lack for the most part the time and inclination to grapple with the important issues of force planning. Instead, their focus tends to be on a particular weapon such as the MX, or the headlines force them to engage in what the military calls fire-fighting. Small changes on the margin of the administration's plan are the most they are likely to propose.

The Congressional Budget Office, established by the Budget Control and Impoundment Act of 1974, has the skills and resources to assist the key congressional committees in forcing the Defense Department to justify its plans and programs as an integrated whole more convincingly than it now does, and to propose alternatives. But because the office must be, and be seen, as nonpartisan in character, it has to be wary—as does the Office of Management and Budget in the executive branch—of inviting the explosive debates that can arise over objectives, forces, and the allocation of resources among missions and services. To suggest, for example, that the 600-ship navy may be a luxury rather than a necessity would arouse the wrath of both congressional constituents and the Navy and risk the loss of access to information that only the Defense Department can provide. The more circumspect course is to underline the heavy mortgage on future resources that a 600-ship navy will entail and to leave implicit the issue of what should be the size and composition of the fleet. Under these conditions the combat force and its relatives, whatever their merits, are unlikely to find sponsors in Congress.

CHAPTER SEVEN

WHAT CAN BE DONE

IT MAY BE too optimistic to suggest that the prospects are good for developing a defense that gives reasonable confidence of maintaining the conditions of U.S. security and at the same time releases resources for other purposes such as the reduction of the deficit. Even after five years of feasting, real famine is unlikely to set in for the Defense Department. Still, the real rate of growth in the defense budget seems almost certain to slow down during the next few years, barring a turn for the worse in international relations. Thus, not inconsiderable resources could be applied to the reduction of the budget deficit. Indeed, as much as 20 percent of the burden could be removed by sensible restraint in the growth of the defense budget.

What remains uncertain is how the slowdown will affect the various appropriation accounts in the defense budget. The case for cutting back primarily on major procurement and some of the more extravagant programs in research and development (such as the Strategic Defense Initiative and the advanced tactical fighter) is strong. But the likelier outcome—especially since being prodefense has become equated with voting for the Strategic Defense Initiative, the MX, Midgetman, and carriers—is that the boom in procurement will continue and the cuts will come in those mundane accounts that make the forces deployable, effective, and sustainable. In fact, no one should be surprised if, in 1992, the programmed force turns out to be more modern and a lot more expensive than the baseline force of 1981, and only marginally better in performance.[1]

1. For a discussion of future defense budget options, given that so much of what has happened since 1981 can no longer be undone, see William W. Kaufmann, *The 1985 Defense Budget* (Brookings, 1984) and *The 1986 Defense Budget* (Brookings, 1985).

The Competition

These may be the realistic prospects for the immediate future, but they need not be inevitable. A great deal can be done to ensure progress toward a more reasonable defense in the next five years despite the unenviable record of the recent past. Perhaps the first condition of progress is a recognition of the competitive nature of the relationship with the USSR and acknowledgment of where the United States stands in the competition.

It would be foolish to pretend that such a competition does not exist or that the military realm can somehow be made exempt from it. It would be equally foolish to equate the current process with the events that preceded World War II when the Western democracies responded with apathy to German and Japanese military dynamism. The pace of the competition since World War II has been, at best, erratic; at no time has either side matched the pace and magnitude of Hitler's buildup in the 1930s. During the 1950s and 1960s, the United States probably invested more in defense than the USSR did, for the last fifteen years, the Soviet investment has undoubtedly been the larger of the two. Over the entire forty years, the real cumulative amounts seem to have been about the same measured in constant dollars.

Future Growth

Whether the competition has entered a new and more dynamic phase remains uncertain. Certainly Brezhnev, with the adoption of what appears to have been a policy of gradual but steady modernization and expansion of conventional and nuclear forces, brought about a major restoration of Soviet arms. The new leadership may well continue it despite the slowdown between 1976 and 1983 and the continued weakness of the Soviet economy. In the circumstances, it has probably become impossible for the United States any longer to satisfy its security commitments with constant real defense budgets. As has often been remarked, the United States has used up the credits and the advantages gained from victory in World War II, an undamaged economy, and a brief nuclear monopoly. World War II is a memory; other economies, including that of the USSR, have been repaired; a nuclear stalemate has become a fact of life. To ensure that future Soviet military improvements are counterbalanced, future real increases in the range of 3 percent a year (on the average) will probably have to be provided to the Defense Department, assuming a one-time reduction in unspent prior-year and current budget authority to dispose of redundant and inefficient programs, an increased efficiency in the use of future resources, and a clearer definition of acceptable objectives and risks.

If increases of this order could be reasonably assured, starting even as late as fiscal 1988, the services would probably adapt to them easily and comfortably. After all, prior to 1981, real growth of 3 percent a year—which allows for a systematic replacement of inventory and the operation of the next generation of more costly capital goods—struck them as an attractive proposition. However, the services did not believe then, and they probably do not believe now, that a steady policy of modest real growth would replace the feast and famine of the past. Consequently, during the last five years, when the Reagan administration opened wide the gates of the Treasury, the services have tried to rush as many programs as possible into the funding stream both to take advantage of the feast while it lasted and to make future cancellations more difficult once the famine returned. A more restrained appetite by the services would almost certainly accompany a period of gradual but sustained real growth. In the process, a great deal of waste could be avoided.

The Defense Intelligence Agency

Real increases in future defense budgets will not by themselves ensure the rational use of available resources. For that to happen the secretary of defense must take the lead in force planning. The system and methodology for achieving sufficient and efficient forces already exist, although both can be improved. The Defense Intelligence Agency, in particular, cries out for reform. When established as a consolidation of the intelligence agencies of the three services, it was supposed to report to the secretary of defense and provide analyses and estimates for the Defense Department as a whole. The Joint Chiefs of Staff, however, opposed that relationship. As a consequence, not only does the agency report to them; it also provides a view of current and future Soviet capabilities that justifies the more extreme demands of the services. Because of this propensity, most secretaries of defense during the last twenty-five years have encouraged the Central Intelligence Agency to undertake competing analyses and estimates. To a degree, the competition has proved fruitful, but the two agencies have spent an inordinate amount of time quarreling with one another, largely because the Central Intelligence Agency has felt obliged to challenge the inflated estimates provided by the Defense Intelligence Agency.

One way to end this unseemly and wasteful squabbling would be to convert the Defense Intelligence Agency into an organization directly responsible and responsive to the secretary of defense. While this would deprive the Joint Chiefs of Staff of authority over the Defense Intelligence Agency, it would not deprive them of its product. The Joint Chiefs would, in any event, still be able to draw on the intelligence organizations of the three services, all of which continue to exist despite the creation of the Defense Intelligence Agency.

Net Assessment

Another organization that deserves reexamination is the office of net assessment. Orginally it reported directly to the secretary of defense; now it is responsible to the under secretary of defense for policy. Establishment of this office was first proposed to the Fitzhugh Blue Ribbon Panel in 1970. At that time the office of systems analysis (now relabeled program analysis and evaluation) tended to be engaged mostly in relatively microscopic cost-effectiveness comparisons. Little was being done to help the secretary of defense determine how successful U.S. forces would be in reaching their planning objectives or in advancing the methodology for making these assessments. Yet relevant plans and programs for the future depended heavily on estimates of U.S. (and allied) strengths and weaknesses in responding to the main planning contingencies. Systems analysis was obviously capable of performing these assessments and had played a big role, along with the services, in the defense review undertaken by the incoming Nixon administration in 1969. However, there was a widespread feeling, not confined to the services, that systems analysis would be biased in its assessments because of its role in evaluating and recommending programs. A separate office without the history and burdens of systems analysis seemed the better way to produce these assessments for the secretary, and to make them regularly rather than at the outset of a new administration.

That expectation, unfortunately, has not been fulfilled. The office dabbles in intelligence issues, makes perfunctory comparisons of U.S., allied, and opposing forces based largely on Defense Intelligence Agency order of battle data, and dispenses funds for research on a variety of undoubtedly worthy causes distantly related to the improvement of net assessments as they were originally conceived. To a large extent this muted and secondary role has evolved because the services and the Joint Chiefs of Staff claim net assessment as their exclusive prerogative and even go so far as to insist that the Central Intelligence Agency forgo any analysis of how Soviet forces are changing in comparison with U.S. and allied capabilities. Such bureaucratic maneuvers are a commonplace of the Washington scene. But a monopoly of an activity as important to the duties of the secretary of defense as net assessment should not be tolerated. Either the secretary should insist that the office of net assessment perform the function for which it was originally created, or he should abolish the office and transfer the function to the office of program analysis and evaluation.

Policy

While he is at it, the secretary might also consider whether the under secretary for policy hinders more than helps the force planning process. One penalty paid for the concentration of so much authority in the hands of the secretary is that an enormous number of issues come to him either directly or because of appeal from a lesser authority. To lighten this burden, the secretary of defense has delegated much authority to assistant secretaries and has traditionally left to the deputy secretary the oversight of the main procurement programs and the widespread intelligence activities of the department. Until recently only the secretary made decisions about plans, programs, and budgets and relationships with the White House. Most of the support for his role in the PPBS came from the Joint Chiefs of Staff and the office of program analysis and evaluation. However, with the creation of the under secretary for policy the essential relationships among objectives, forces, and budgets have been fractured. In a throwback to standard military planning practice—

with budgets subordinated to "requirements"—the under secretary for policy is responsible for drafting the defense guidance, while other offices deal separately with forces and costs.

Analysis

Arguably, the office of program analysis and evaluation, which began as part of the Comptroller's office, should have stayed there or been returned to the budgetary fold once it became such a target of military and congressional hostility. To fracture the connections among plans, programs, and budgets still further with the creation of the under secretary of policy has been to risk returning the force planning process to the 1950s, when the programs of the individual services were so divorced from budgetary realities that continuity in planning became impossible. Under those conditions, a multiplicity of projects that started as items of fairly small cost soon ballooned more rapidly than growth in the budget could accommodate. The result, inevitably, was cancellations with nothing to show for the investments, or the uneconomic stretch-out of programs with higher costs for each unit of output.

That profligacy was what the linkages among planning, programming, and budgeting helped to bring under control. The linkages are surely worth preserving. The best way to do so is by returning the issuance of the defense guidance to the office of program analysis and evaluation and returning that office to the Office of the Comptroller. Since the only other function of the under secretary for policy is the supervision of the old office of international security affairs (the Defense Department's state department now arbitrarily split in two), which already has assistant secretaries to do just that, the disappearance of the under secretary would not be noticed. The elimination of the bureaucracy and the committees that have grown up around the drafting of the defense guidance would also be appropriate.

Even stripped of these frills, the PPBS is unwieldy. It carries an excessive weight of committees and has become more a vehicle for achieving a consensus than a process for developing and clarifying the hard choices that are necessary in the most prosperous times and essential when budgets grow restrictive. Participation in its deliberations could undoubtedly be cut by 50 percent without any loss of representation or consensus, although individual egos might suffer. Further cuts would damage all three, but would probably improve rationality, a trade well worth considering.

Changes in the analytical community are needed as well. To the extent that the community is allowed to function at all, it tends to define issues to make them amenable to existing methodologies. This, in turn, means a concentration on narrowly conceived comparisons of weapons systems with well-defined costs and measures of effectiveness, to the neglect of the more difficult issues of sufficient and efficient forces that must be the primary concern of the secretary of defense. In the past the office of program analysis and evaluation found it necessary to go beyond the organization of analysis by mission and establish branches that would look at planning contingencies such as attacks on Central Europe, South Korea, and the oil states of the Persian Gulf. The office could do with a modest increase in staff to draft the defense guidance, make net assessments (if the office of net assessment is abolished), and provide the secretary with illustrative alternative forces and budgets. The more microscopic analyses would continue, but the larger issues would receive the attention they deserve.

The Joint Chiefs of Staff and the Secretary of Defense

The Joint Chiefs of Staff and its organization, along with the three services, will inevitably continue to be heavily involved in the force planning process. Their participation would undoubtedly be improved if the chairman had a genuine staff of his own made up, not of envoys from the services (disguised as a joint staff), but of officers whose primary loyalty is to the office of the chairman. In the final analysis, however, the system can work only if the secretary of defense takes his responsibilities seriously. He already has the authority to resolve disputes among the services. He can obtain the knowledge and the skills necessary to set realistic, affordable objectives and make reasonable choices at minimum cost to the taxpayer. But the secretary must command more than good intentions. He must develop the instincts of an analyst himself.

Caspar W. Weinberger, the current secretary of defense, has written that "we are perhaps in danger of becoming a nation of ascetic systems analysts, without the glowing fire and the vision and the ability to inspire that Churchill possessed in such full measure."[2] The United States may also be in danger of forgetting that visions, if they are to be anything more than that, have to be implemented with specific objectives, forces, and resources. It was Winston Churchill who declared: "You cannot ask us to take sides against arithmetic. You cannot ask us to take sides against the obvious facts of the situation."[3] The arithmetic remains to be learned.

2. Quoted in *Time* (February 11, 1985), p. 29.
3. Quoted in *Department of Defense Annual Report, Fiscal Year 1981*, p. 1.